Deep Interior of the Earth

Topics in the Earth Sciences

SERIES EDITOR

T.H. van Andel
University of Cambridge

Titles available

Deep Interior of the Earth

J.A. Jacobs

Honorary Professor Institute of Earth Studies, University College of Wales Aberystwyth

and

Honorary Professor of Geophysics University of Cambridge

CHAPMAN & HALL
London · New York · Tokyo · Melbourne · Madras

Published by Chapman & Hall, 2-6 Boundary Row, London SE1 8HN

Chapman & Hall, 2-6 Boundary Row, London SE1 8HN, UK

Blackie Academic & Professional, Wester Cleddens Road, Bishopbriggs, Glasgow G64 2NZ, UK

Chapman & Hall, 29 West 35th Street, New York NY10001, USA

Chapman & Hall Japan, Thomson Publishing Japan, Hirakawacho Nemoto Building, 6F, 1-7-11 Hirakawa-cho, Chiyoda-ku, Tokyo 102, Japan

Chapman & Hall Australia, Thomas Nelson Australia, 102 Dodds Street, South Melbourne, Victoria 3205, Australia

Chapman & Hall India, R. Seshadri, 32 Second Main Road, CIT East, Madras 600 035, India

First edition 1992
Reprinted 1993

© 1992 J.A. Jacobs

Typeset in 11/12 Bembo by Best-set Typesetter Ltd., Hong Kong

ISBN 0 412 36570 7

A catalogue record for this book is available from the British Library
Library of Congress Cataloging-in-Publication Data available

Contents

Series foreword

Year by year the Earth sciences grow more diverse, with an inevitable increase in the degree to which rampant specialization isolates the practitioners of an ever larger number of subfields. An increasing emphasis on sophisticated mathematics, physics and chemistry as well as the use of advanced technology have set up barriers often impenetrable to the uninitiated. Ironically, the potential value of many specialities for other, often non-contiguous ones has also increased. What is at the present time quiet, unseen work in a remote corner of our discipline, may tomorrow enhance, even revitalize some entirely different area.

The rising flood of research reports has drastically cut the time we have available for free reading. The enormous proliferation of journals expressly aimed at small, select audiences has raised the threshold of access to a large part of the literature so much that many of us are unable to cross it.

This, most would agree, is not only unfortunate but downright dangerous, limiting by sheer bulk of paper or difficulty of comprehension, the flow of information across the Earth sciences because, after all it is just one earth that we all study, and cross fertilization is the key to progress. If one knows where to obtain much needed data or inspiration, no effort is too great. It is when we remain unaware of its existence (perhaps even in the office next door) that stagnation soon sets in.

This series attempts to balance, at least to some degree, the growing deficit in the exchange of knowledge. The concise, modestly demanding books, thorough but easily read and referenced only to a level that permits more advanced pursuit will, we hope, introduce many of us to the varied interests and insights in the Earth of many others.

The series, of which the book forms a part, does not have a strict plan. The emergence and identification of timely subjects and the availability of thoughtful authors, guide more than design the list and order of topics. May they over the years break a path for us to new or little-known territories in the Earth sciences without doubting our intelligence, insulting our erudition or demanding excessive effort.

Tjeerd H. van Andel
Series Editor

Preface

This book is concerned with the deeper regions of the Earth and what part, if any, processes there may play in events at the surface of the Earth. Our geological thinking has been revolutionized over the last few decades by the theory of plate tectonics which embodies the older ideas of continental drift. Although the broad concepts of plate tectonics are now generally accepted, there is still some controversy over what drives the plates. Attention has usually been focused on the upper few hundred kilometres of the Earth, but it is now appreciated that the deeper parts of the Earth may also be involved. It is still being debated whether convection in the mantle is two-stage with a boundary at the seismic discontinuity at a depth of about 670 km, or whether it consists of a single convection cell. Moreover it is not universally agreed whether the seismic discontinuity at 670 km is a compositional or a phase change. Another question is the possible existence of thermal mantle plumes and whether they extend from the core-mantle boundary right up to the lithosphere. Reversals of the Earth's magnetic field caused by changes in the Earth's core have also played a key role in the development of the theory of plate tectonics.

The boundary between the Earth's core and mantle at a depth of about 2900 km has aroused particular interest in recent years. The outer core consists of liquid iron alloyed with one or more lighter elements, whilst the overlying mantle is a solid silicate, most probably at a much lower temperature. The interaction between two such very different materials could be thermal, chemical or mechanical and could have profound effects on mantle convection. The topography, thermal and chemical conditions at the core mantle boundary could also influence the flow in the fluid outer core which is responsible for the dynamo that generates the Earth's magnetic field. Other theories maintain that convection in the outer core is driven by heavier material freezing out to form the solid inner core.

To learn about the Earth's deep interior, many different disciplines are involved. This book discusses the origin of the Earth and the formation of the core, the chemical composition and physical properties of the deep Earth and the origin and maintenance of its magnetic field. The Earth is also considered as a member of the solar system and the internal constitution of the other terrestrial planets and the Moon briefly discussed.

__1__

The origin of the Earth

1.1 INTRODUCTION

This book is about the Earth's core – the innermost regions of the Earth. A cross section of the Earth shows that there is an outer shell, called the **crust** which is only a few tens of kilometres thick beneath the continents and about 5 km beneath the oceans. Below the crust there is a rocky **mantle** extending to a depth of about 2900 km – just under halfway to the centre of the Earth. Below the mantle is the **core** which is mainly iron and which consists of an outer core (OC) that is fluid and an inner core (IC) that is solid (Figure 1.1). The radius of the IC is about one fifth that of the Earth. How we know these facts and other information about the core will be described in the following chapters. We shall see that the core plays an important role in many geophysical studies, e.g., it is motions in the fluid OC that generates the Earth's magnetic field. The IC, in spite of its small size (its volume is less than one per cent that of the whole Earth) is important. Present estimates of the temperature at the centre of the Earth, of the energy sources that are required to sustain its magnetic field and of the heat from the core that is available to drive convection in the mantle all depend on the physical and chemical state of the IC.

Our knowledge of the interior of the Earth is pieced together from many disciplines – physics, chemistry, geology and astronomy. It is thus difficult for one person to keep abreast of new developments in a number of different fields and to appreciate their possible significance for the Earth sciences. The rapid advances in computer science have meant that vast amounts of data can now be analysed to construct an Earth model or test some theoretical idea, e.g., over a million seismic observations have been used to model the structure of the Earth's interior (Dziewonski and Woodhouse, 1987).

The Earth has undergone continual change throughout its history, but its evolution has been very different from that of the other terrestrial planets and the Moon. Our knowledge of these bodies has been greatly extended in the last thirty years through

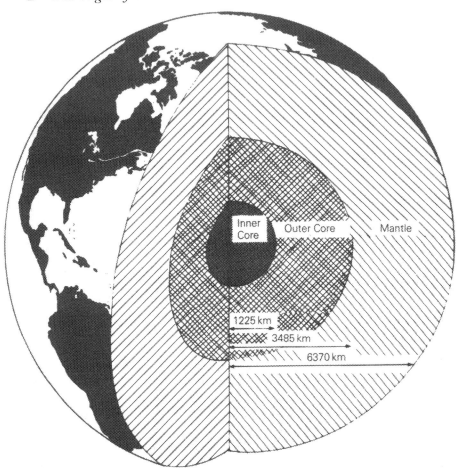

Figure 1.1 Cross-section of the interior of the Earth. On this scale the crust cannot be seen.

space exploration. Mercury, nearest to the sun, still retains the record of intense meteorite bombardment during the first half billion years or so of its existence on its surface. Thus, unlike the Earth, Mercury is seen to have been tectonically quiet for the past two to three billion years.

Venus is often referred to as the Earth's twin, having comparable mass, radius, density and gravity. However, it is blanketed by a thick, heavy atmosphere of CO_2 which insulates the surface leading to temperatures of $\sim450°C$. With the present controversy of a possible future greenhouse effect on Earth, one can but wonder whether Venus once had an ocean which boiled away under runaway greenhouse conditions. Alternatively, Venus being much closer to the sun, perhaps was too hot for water vapour to condense into oceans and CO_2 to be locked up in surface rocks. Because of the dense cloud cover, radar imaging instead of normal photography must be used to 'see' the surface of the planet. The first planet-wide data on the surface topography were obtained by the Pioneer Venus Orbiter in 1978. However the horizontal resolution was only 100–200 km. The resolution was considerably improved in 1983 and 1984 by the Soviet spacecraft Venera 15 and 16 which obtained radar images with a horizontal resolution of 1–2 km. A wide variety and complexity of surface features were revealed – volcanoes, mountain ranges and impact craters. One mountain range is over 11 km high, higher than Mount Everest. There is also evidence of horizontal motion of the crust of Venus, including regions of both extension (rifting) and compression (ridges and mountains). Our present knowledge of the geology and tectonics of Venus has been reviewed by Head and Crumpler (1990). Because there is no water on Venus, there is no erosion so that any tectonic activity – mountain building, faulting, volcanism – is not destroyed. One outstanding question is whether the surface of Venus has a system of plate tectonics like that on Earth. Most geophysicists think not, although Head is not convinced that this is the case and is of the opinion that some plate tectonic processes exist on Venus similar to those on Earth.

It is to be hoped that the answers to many of these questions will be answered by the Magellan spacecraft which went into orbit around Venus on Aug 10, 1990. A special section of Geophysical Research Letters (vol. 17, Aug 1990) was devoted to highlight some of the issues that it is hoped the Magellan mission will settle. Magellan is orbiting Venus in a near polar orbit at an altitude that varies between 290 and 8500 km. It orbits about 8 times every 24 hours and each orbit images a section approximately 20 km wide and 7000 km long. It was originally planned that in the first eight months of continuous radar and altimetry mapping, Magellan would cover nearly 90% of the surface of Venus at a resolution of a few hundred metres (Saunders *et al.*, 1990). Additional cycles would complete global coverage and enable a high resolution gravity map to be drawn. Unfortunately soon after Magellan began returning images of the Venusian terrain with a resolution of 120 m, troubles developed in the spacecraft's control systems. On two occasions, radio contact with Magellan was lost – on Aug 16 for 14 hours and on Aug 21 for 18 hours. These problems were quickly solved and in its first 243 cycles (one Venus-year) 84% of the planet's surface was surveyed. The surface of Venus is cross-hatched with faults, long mountain ranges, and lava flows – one extinct lava channel is more than 200 km long. Domes 3–10 km across dot the planes – one flat-topped dome has been described by Head as an 'upside down cereal bowl'. Some of the surface features are similar to those found on Earth, such as the East African Rift Valley, the San Andreas fault off California and the volcanoes on Hawaii. There are also many impact craters – most of them are rather large, probably because the planet's atmosphere is so thick that all but the largest meteorites would be burnt up before they could reach the surface. One crater, Golubkina, is 34 km across and has terraced inner walls and a central peak, typical of large impact features on the Earth, the Moon and Mars. Further imaging of Venus has revealed four large areas ranging in size from a few to 5 million km^2 that show no impact craters at all – despite the bombardment to which the surface must have been exposed. This would seem to imply that, at least in some regions, there has been recent geological activity which has obliterated any craters by lava flows or by tectonic movements. A detailed analysis of the initial data obtained from Magellan has been given in *Science* **252**, 247–312 (1991). No Venusian plate tectonics appears to have been observed. The lack of any detectable magnetic field raises questions about the state of any Venusian core – this will be discussed in section 7.4.

The surface of Mars, like that of Mercury, retains a record of intense meteorite bombardment. The thousands of photographs of its surface also show evidence of volcanism, tectonic activity and erosion by wind and running water. One volcano (Olympus Mons) is over 600 km wide at the base, 27 km high with a summit caldera 80 km wide. There is also evidence of catastrophic flooding, although there are no signs of ancient seas or accumulation of sediments. An interesting question is 'where is the water now?'. Presumably it must be trapped underground or in aquifers or exist as ice, since the present CO_2 atmosphere has too low a pressure for liquid water to exist – it would vaporize. There are also deep canyons and rifts. One canyon (Valles Marineris) is ~2700 km long, in places as wide as 500 km, and 6 km deep. Presumably it is of tectonic origin. Again the lack of any detectable magnetic field raises the question of a Martian core – this will be discussed in section 7.3.

Finally the surface of the Moon also bears witness to early meteorite bombardment. Lacking an atmosphere and water, the craters, unlike those on the Earth, remain to-day, except where they have been covered by later impacts. In addition lunar samples brought back by the Apollo spacecraft have shown that the lunar basins (the maria) are flooded by basaltic lava, which flowed out from the interior of the Moon following the formation of the basins by large impacts. Unlike volcanic activity on the Earth, which still continues to-day, the youngest lavas on the Moon are ~3000 Ma old. Thus the Moon, though initially geologically active, is now a dead planet and has been so for ~3000 Ma. A better understanding of the Earth will only come when we know why the different planets have evolved so differently.

We shall discuss later the differentiation of the Earth into crust, mantle and core. A question we may well ask is, when did this differentiation take place? Has the Earth always had a core, or has its present structure evolved over geologic time? This question cannot be divorced from the broader question of the origin of the Earth itself and indeed the origin of the solar system. On an even larger scale, there are more than 200 000 million stars in our galaxy (the Milky Way) alone, and it is unreasonable to suppose that our solar system is unique. It is far more likely that planets are normally present in the vicinity of certain stars like our sun. In attempting to answer such cosmological questions, it must be stressed that it is by no means certain that we can give definite answers to many of the problems. It may well be that all record of the circumstances under which our solar system came into being has now been lost. We will discuss in the next two sections the age of the Earth and the origin of the solar system.

1.2 THE AGE OF THE EARTH

Until the discovery of radioactivity, the age of the Earth was a matter of much debate. The phenomenon of radioactivity was discovered by Becquerel in 1896, although it was not until 1905 that Rutherford suggested that radioactive minerals could be used to date rocks. Most naturally occurring elements have isotopes with stable nuclei, although there are several with unstable nuclei. The atoms of these radioactive elements (such as ^{238}U, ^{235}U, ^{232}Th and ^{40}K) spontaneously decay by emitting atomic particles (helium nuclei and electrons).

The average rate of disintegration is constant and does not vary with any changes in chemical or physical conditions that affect most chemical or physical processes. What we can measure is the time that has elapsed since the radioactive parent became incorporated into a rock in which their daughter elements were trapped. Successful measurements had to await the development of the mass-spectrometer. This instrument produces a beam of electrically charged atoms from the rock which is to be dated. The beam is passed through electric and magnetic fields that deflect the atoms by an amount that depends upon their masses. Thus isotopes of elements can be separated.

The fundamental equation of radioactive decay was shown by Rutherford in 1900 to be

$$\frac{dP}{dt} = -\lambda P \qquad (1.1)$$

where P is the number of parent atoms at time t, and λ the decay constant. Integration of equation (1.1) gives

$$P = P_o e^{-\lambda t} \qquad (1.2)$$

where P_o is the number of parent atoms at time $t = 0$. Solving for t, equation (1.2) gives

$$t = \frac{1}{\lambda} \ln \left(\frac{P_o}{P} \right) = \frac{1}{\lambda} \ln \left(1 + \frac{D}{P} \right) \qquad (1.3)$$

where $D = P_o - P$ i.e., the number of daughter atoms at time t.

Radioactive decay rates are often expressed in terms of the half-life, $T_{1/2}$, which is the time in which one half of the radioactive nucleus will decay. It follows from equation (1.2) that

$$T_{1/2} = \frac{1}{\lambda} \ln 2 \simeq \frac{0.693}{\lambda} \qquad (1.4)$$

The isotopes most used for age determination, together with their half lives are given in Table 1.1.

It can be seen that ^{40}K decays in two ways. About 89% of the ^{40}K atoms decay to ^{40}Ca, and 11% to ^{40}Ar. The decay to ^{40}Ar is that used for dating, since ^{40}Ar can be easily distinguished from ordinary argon formed in other ways, whereas ^{40}Ca is ordinary calcium and atoms of radiogenic origin cannot be distinguished from others.

There are other long lived radioactive nuclides, many with half-lives greater than 10^{11} y. It is noteworthy that all radioactive

Table 1.1. Some long-lived radioactive nuclides

Nuclide	Stable products	Half-life (years)
^{40}K	^{40}Ca, ^{40}Ar	1.3×10^9
^{87}Rb	^{87}Sr	4.9×10^{10}
^{147}Sm	^{143}Nd	1.1×10^{11}
^{187}Re	^{187}Os	4×10^{10}
^{232}Th	^{208}Pb	1.4×10^{10}
^{235}U	^{207}Pb	7.0×10^8
^{238}U	^{206}Pb	4.5×10^9

nuclides present on the Earth in significant amounts have half-lives longer than about 10^9 y. This gives an order of magnitude estimate of the age of the Earth, assuming that shorter-lived radioactive nuclides that were originally present have long since decayed. A minimum age of the Earth is given by the age of the oldest terrestrial rocks – rocks 3900 Ma old have now been found and grains of zircon as old as 4300 Ma, although not the rocks from which the grains came. Moreover all meteorites, regardless of their composition, are of the same age (~4500 Ma). This suggests that all meteorites originated in other bodies of the solar system that formed at the same time as the Earth.

Further information on the early history of the Earth can be obtained from xenon isotopes. ^{129}I decays to ^{129}Xe with a half-life of approximately 17 Ma and ^{128}Xe is formed in the spontaneous fission of ^{244}Pu (half-life ~82 Ma). Excess ^{129}Xe and ^{128}Xe have been found in some meteorites, suggesting that element synthesis took place some 200 Ma before the solar system was formed. Excess ^{129}Xe has also been found in some terrestrial samples – from mid-ocean ridge basalts, mantle xenoliths and from gas from a CO_2 well in New Mexico. Ozima *et al.* (1985) have reviewed the data on the Xe isotopic composition of terrestrial samples and concluded that the Earth's inner regions accreted a few tens of millions of years earlier than the outer regions from which the atmosphere evolved from degassing of the mantle.

1.3 THE ORIGIN OF THE SOLAR SYSTEM

There are a number of possibilities of the way in which the Earth and other members of the solar system could have formed. They may have cooled from a hot gas or accreted cold from dust particles in the solar nebula. The proto-Earth may have been homogeneous, the present differentiation into core, mantle and crust taking place later. Alternatively the mainly iron core of the Earth may have formed first, the silicate mantle being deposited upon it later. Inhomogeneous models like this could also have either a hot or a cold origin. A hot origin involves a knowledge of the order of condensation of elements in the primitive solar nebula; a cold origin involves a knowledge of the mechanism of the accretion process. Although attractive in some ways, inhomogeneous models are now generally out of favour. The temperatures for the condensation of iron and of the major silicate phases from the solar nebula are so similar (to within a few tens of degrees) that efficient separation at that stage is extremely unlikely. In the absence of an efficient physical or chemical mechanism for such a large-scale separation it is more likely that the Earth accreted in an approximately homogeneous fashion.

The evolution of the planets from the solar nebula can be considered in three main stages – an early stage producing kilometre-sized planetesimals formed by small-scale gravitational instabilities, an intermediate stage during which kilometre-sized planetesimals grow to 1000 km sized bodies, and a final stage during which 1000 km-sized bodies grow to full sized planets (e.g., Wetherill, 1980). The evolution of the second stage would be governed by two competing mechanisms – collisional fragmentation and gravitational accumulation. We shall confine ourselves to the final stages of the growth of a terrestrial planet which would affect many of its fundamental properties such as its thermal history and petrology.

Wetherill (1985) carried-out computer models of the accretion process (Figure 1.2). He found that most of the mass would tend to be concentrated in larger bodies, resulting in about 500 planetesimals of about one-third the moon's mass. A few would then grow at the expense of the rest and, once a proto-Earth was formed, giant impacts would be inevitable. Collisions between proto-planets have been invoked to explain some of the anomalies in the present day solar system, such as the retrograde rotation of Venus, the exceptional tilt of the rotational axis of Uranus and the very high density of Mercury. The case of Mercury and the Earth-Moon system will be considered in the next two sections.

Further evidence of large impacts in the solar system comes from some meteorites found on the Earth. Many thousand meteorites have been found in Antarctica, of which a few are believed to have been ejected from the Moon following a lunar impact event (e.g., Eugster, 1989). One of the main reasons for this suggestion is their close resemblance to lunar samples brought back by the Apollo spacecraft. It has been further suggested that some meteorites have come from Mars. One of the reasons for this is the discovery of glassy nodules in a meteorite which contain a noble gas component whose elemental and isotopic composition of Ne, Ar, Kr and Xe does not resemble that trapped in other meteorites, but does resemble in great detail the compositional pattern for these gases in the Martian atmosphere as measured by the Viking spacecraft. Such an agreement has also been found for the isotopically heavy ^{15}N and CO_2. If some meteorites come from Mars, detailed analysis should reveal many properties of Mars as a whole. It would seem on the basis of the Rb-Sr isotopic composition that the core of Mars formed about 4500 Ma ago. Moreover the depletion of sulphur-loving elements suggests that the core is rich in sulphides, sulphur constituting about 14% of the mass of the core (also section 7.3). A

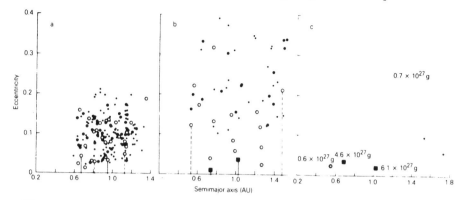

Figure 1.2 Results of numerical simulation of the accumulation of 500 bodies. The eccentricity of their orbits is plotted against the semi-major axis measured in AU. (1 AU is the mean distance of the Earth from the sun). (a) After 1.8 Ma, 18% of the mass (~60% of the final radius) has accumulated in the final planets. Small planet-sized bodies (10^{26} to 10^{27} g shown as large open circles), have grown as well as bodies in the lunar mass range (2.5×10^{25} to 10^{26} g shown as large filled circles). A number of small bodies (2.5×10^{25} g) shown as small filled circles have not accreted. (b) After 9 Ma, 66% of the mass (~87% of the final radius) has accumulated in the final planets. Two bodies (shown as filled squares) are larger than 10^{27} g and a number of massive bodies are in rather eccentric orbits. (c) After 252 Ma, almost all the bodies have been accumulated into planets with masses and orbits resembling the present terrestrial planets. A few small Mars-crossing bodies remain and are removed by ejection from the solar system, being trapped into asteroidal orbits or planetary collision during the following 250 Ma (after Wetherill, 1985).

good review of the formation of the Earth has been given by Wetherill (1990).

1.3.1 Mercury

Mercury is less than one tenth as massive as either Venus or the Earth, but has an exceptionally high density – its Fe/Si ratio approximates twice that of the other terrestrial planets. The 'conventional' explanation for this has been the chemical fractionation of material that originally condensed out of the solar nebula. If the temperature of the nebula varied inversely with distance from the proto-sun, equilibrium condensation would favour Fe-rich condensates at Mercury's orbit, which is closest to the sun. Alternatively if the comparatively small mass of Mercury is the result of mass loss by small particles spiralling into the sun, low density Si particles would be lost from Mercury's orbit before its accumulation. Both these models require that Mercury's distance from the sun remains isolated from the bulk of the material that eventually went to form the terrestrial planets. However, Wetherill's (1989) computer calculations showed that once planetesimals reach lunar size, they would gravitationally perturb each other into crossing orbits. Thus a typical body's orbit is likely to wander randomly over much of the area now occupied by the terrestrial planets, thereby reducing the chemical fractionation associated with the heliocentric distance at which the planetesimals were originally formed.

Benz *et al.* (1988) have carried out numerical simulations of the effect of planetary-scale collisions. In the case of a giant collision of a body with the proto-Mercury, much of Mercury's silicate mantle would escape from the planet's gravitational field, leaving behind an Fe-rich core. If the proto-Mercury was in orbit crossing the orbits of the much larger proto-Venus or proto-Earth at the time it experienced such a catastrophic collision, much of the ejected material would be accreted by Venus and the Earth.

1.3.2 The Earth-Moon System

There has been much speculation about the origin of the Moon. The three main contenders – fission from the Earth, capture by the Earth and co-accretion alongside the Earth – all have problems. The main difficulty has been to account for the large amount of angular momentum of the Earth's rotation and the Moon's revolution about the Earth. Since 1984 attention has been focused on a fourth possibility – the collision of the partially formed proto-Earth with a body about the size of Mars. This was first suggested by Cameron and Ward in 1976. Later computer calculations by Wetherill (1985) and others demonstrated the feasibility of such a scheme – it can drive a significant amount of material into orbit to form the Moon and satisfy the angular momentum requirements. Newson and Taylor (1989) have also considered the consequences of such an event for the bulk chemistry of the Earth and Moon and concluded that the hypothesis is cosmochemically plausible.

However, there is one problem with such a scenario. Because of the enormous energy involved in such a collision, one would expect to see some record of it preserved in the Earth's present structure. Calculations by many workers indicate that the impact would have completely melted the proto-Earth, yet some geochemists claim that the upper few hundred kilometres of the mantle show no evidence of any such melting. Whenever crystals separate from a melt, either during melting or crystallization, a characteristic record is left on the chemistry of the melt and crystals. This is because some elements (the incompatible or trace elements) fractionate preferentially into the melt, whereas others prefer the solid phases. A melt can crystallize to a solid of exactly the same composition only if the crystals remain in continuous contact with the melt. In other words if the early Earth once had a 'magma ocean', it would not freeze like an ice cube – some minerals would freeze out first and either sink to the bottom or rise to form a surface scum. This would alter the composition of the remaining magma and thus the composition of minerals that form later. Ringwood (1989) has looked at a number of elements whose relative abundances should have been noticeably altered in the first crust to have formed on a molten Earth, and found no evidence of such alteration in the oldest minerals known (4200 Ma old zircons). Ringwood thus concluded that there never was a terrestrial magma ocean and therefore no giant impact. Moreover the mechanics of the giant impact theory imply that the Moon is derived mainly from the mantle of the impactor, and Ringwood (1979) maintains that geochemical evidence strongly indicates that the material of the Moon came mainly from the terrestrial mantle.

The basis of his argument is the lower mean density of the Moon ($3.344 \, \text{g cm}^{-3}$) which is almost the same as that of the Earth's mantle, and the lack of volatile elements in the Moon relative to the Earth as shown by the Apollo missions. Ringwood et al., (1990) also point out the similarity in the abundances of V, Cr and Mn in the mantles of the Earth and Moon. They argue that at the pressures corresponding to those in the Earth's lower mantle, Cr, V and perhaps Mn are preferentially partitioned into molten Fe. This is just the opposite to their behaviour at the lower pressures and temperatures in the Earth's upper mantle and the Moon, where experimental studies of the partitioning of these elements between molten Fe and silicates has shown that they are lithophile.[*] Ringwood et al. thus attribute the depletion of these elements in the Earth's mantle to their siderophile[†] behaviour during formation of the Earth's core at pressures that were sufficiently high to cause substantial amounts of oxygen to dissolve in molten metallic Fe. (section 3.4.2).

There are possible ways around these geochemical arguments, e.g., at great depths within the magma ocean, the residual liquid and the new crystals might not have buoyancies sufficiently

---

[*] Lithophile – having a strong affinity for oxygen concentrated in the silicate minerals.

[†] Siderophile elements are those with weak affinities for oxygen and sulphur but soluble in molten iron.

different to separate them. Chemical fractionation, but not physical separation would thus occur so that bulk rock composition might not reflect melting. Alternatively if crystals could be kept in contact with the melt throughout crystallization, perhaps by entrainment in a turbulently convecting flow, the freezing of a magma ocean would have left no imprint on the chemistry of the upper mantle.

Another way out of these difficulties has been proposed by Ringwood (1989). In his model a giant impact occurred after the Earth had accreted to 70% of its present size, satisfying the high momentum density of the Earth-Moon system. The Moon however was not produced by this impact, but was formed from ejected material from the Earth's mantle following collisions of much smaller, high velocity bodies with the Earth at a later stage in the accretion of the Earth. Such impacts would not cause gross melting and differentiation of the Earth.

Ryder (1990) has carried out a detailed search of the samples recovered by the Apollo and Luna spacecraft and lunar meteorites and found no evidence for a single sample of an impact melt older than 3850 Ma. The data thus indicate only light bombardment in the first 600 Ma and then an intense cataclysmic bombardment. This is consistent with the above suggestion for the origin of the Moon, since it implies late and rapid accretion of the Moon from Earth orbiting material, rather than prolonged accretion from a heliocentric swarm of smaller bodies. Ryder suggests that several moons may have been formed from the material thrown out by the impact of a large Mars-sized body with the Earth. The orbits of two of them may have closed gradually until a collision at 3900 Ma produced the cataclysmic bombardment of the Moon and Earth.

The fact that the Earth has a large moon and a rapid prograde rotation whereas Venus has no moon and rotates slowly retrograde must be a consequence of their formation – possibly resulting from collisions among very large planetesimals in the terminal stage of the formation of the terrestrial planets. Perhaps the proto-Earth was hit by a very large (Mars-sized) body leading to the formation of the Moon, whilst the greatest impact on Venus was much smaller. A review of the giant impact hypothesis for the origin of the Moon has been given by Stevenson (1987).

It must be stressed that the giant impact hypothesis for the evolution of the terrestrial planets has not been 'proved'. It is impossible to establish any hypothesis about the origin of the solar system beyond a shadow of doubt. It has been discussed here because it presently enjoys favour and has important consequences for the differentiation of the Earth into core and mantle.

1.4 'CHAOS' IN THE SOLAR SYSTEM

It was long believed that once the physical laws governing a system were known, the past and future behaviour of the system could be found by solving the equations with the appropriate initial conditions. For example, the motion of a ball in a spinning roulette wheel is a deterministic system – the ball and wheel are subject to known forces, yet it is all but impossible to predict the final outcome in an initial spin. The study of chaos has shown that even completely deterministic systems such as those involving gravitational interactions can be chaotic.

Poincaré was the first to appreciate the problems that could arise from the gravitional interaction of more than two bodies as is the case in the solar system. He also realized that in the solar system chaos and order were closely connected with resonance. Resonance may occur when there is a simple numerical relationship between two periods that leads to repeated configurations. There are many resonances in the solar system, e.g., the Jovian satellite Io has an orbital period which is nearly one half that of the next satellite Europa; they are in a 2:1 orbit-orbit resonance. The situation is even more complicated since Europa is simultaneously in a 2:1 resonance with the next satellite Ganymede. There is also a 3:2 orbit-orbit resonance between Neptune and Pluto. Another common form of resonance in the solar system is spin-orbit resonance, e.g., the 1:1 spin-orbit resonance of the Moon that forces it to keep the same face towards the Earth, and the 3:2 spin-orbit resonance of Mercury. Most natural satellites in the solar system are in synchronous spin states because of tidal effects.

Sussman and Wisdom (1988) integrated the orbits of the outer planets over 845 Ma. They found that Pluto's orbit appears to be chaotic, due in part to its 3:2 orbit-orbit resonance with Neptune. Wisdom *et al.*, had predicted earlier in 1984 that Saturn's satellite Hyperion would be twisting and tumbling chaotically so that it would be impossible to predict its behaviour in any detail. They suggested that the cause of its chaotic rotation was its irregular shape and perturbations of its orbit by the gravitational pull of Saturn's larger satellite Titan. (The orbital periods of Hyperion and Titan have a 4:3 relationship.) Direct observations have now confirmed the computer prediction that the motion of Hyperion is indeed chaotic.

Laskar (1989) has now shown that the motions of the planets of the solar system are chaotic. This does not mean that the Earth could at any moment fly off into intersteller space or crash into the sun, but that it is impossible to predict the long term behaviour of the solar system because any uncertainty in the initial conditions of the system increases exponentially with time: in other words no matter how exactly present planetary motions are known, motions over geologic time cannot be predicted. Laskar has shown that in fact it is not possible to predict the orbits of the inner planets (which include the Earth) for more than a few tens of millions of years – a perturbation as small as 10^{-10} in the initial conditions will lead to 100% discrepancy after 100 Ma. This sensitivity to initial conditions may have determined the general character of the solar system as we see it today. Chaotic behaviour has now been found in many fields of science e.g., chemical reactions, the weather, and population fluctuations of plants and animals.

There is now substantial evidence that deterministic chaos is a plausible mechanism for the observed seismic patterns and earthquake statistics (Huang and Turcotte, 1990). Numerical models of flow in the Earth's mantle suggest that convection may be chaotic. Chemical heterogeneities exist in the mantle as a result of subduction of oceanic lithosphere and possibly by subduction of sediments and disturbance of the continental lithosphere with the entrainment of lower mantle material in upwelling plumes (section 6.5). Such heterogeneities are stirred into the surrounding mantle matrix by convective shear. Chemical equilibrium is finally achieved by diffusion. Numerical models carried out by Kellogg and Turcotte (1990) have shown that particle paths are chaotic both in time and space. Mixing takes place very rapidly – for heterogeneities with an initial size of 6 km the mixing time is 240 Ma for layered mantle convection and 960 Ma for whole mantle convection (section 6.6). Chaotic behaviour may also be a key factor influencing reversals of the Earth's magnetic field (section 5.5).

An excellent introductory account of chaos has been given by Gleick (1988), and by Wisdom (1987) for the existence of chaos in the solar system. A number of papers on chaos have also been published in *Science* (Jan 6, Jan 20, Feb 17, March 10 and July 7, 1989) and a continuing series of papers on chaotic behaviour in various fields in the *New Scientist* (Oct 21, Oct 28, Nov 4, Nov 11, Nov 18, Nov 25, Dec 2, 1989 and March 24, June 9, June 30, Aug 11, Sept 15, Sept 22, Sept 29 and Oct 6, 1990). The article in the Nov 25, 1989 issue deals with chaos in the solar system.

1.5 TIME OF CORE FORMATION

The constitution of the Earth with be discussed in Chapter 3. For the moment we will accept the conclusion that the core is composed mainly of Fe with some 10% light alloying element (probably S or O) overlain by a silicate mantle. There is one constraint on possible models of the Earth. We shall see in section 5.3 that the Earth's magnetic field is most probably the result of motions in the liquid, mainly iron outer core (OC). Rocks as old as 3500 Ma have been found which possess remanent magnetization, so that it is extremely probable that the Earth had a molten OC about the same size as that at present 3500 Ma ago.

Many models of core formation have been proposed and we will not attempt to discuss all of them. Elasser (1963) considered an initially homogeneous Earth that accreted cold. He assumed that as we go deeper and deeper into the Earth, the melting point of silicates increases faster than the actual temperature. This would imply that the viscosity of silicates should increase with increasing depth. Thus as the proto-Earth heats up by radioactivity, the outer layers would be the first to become soft enough to permit iron to sink towards the centre – further down the fall of iron is slowed by increased viscosity. Elsasser then suggested that a layer of iron would form, which, however, would be gravitationally unstable and tend to form large drops. These would fall rapidly to the centre of the Earth forming a proto-core. Elsasser estimated that it took several hundred Ma to form a core roughly the size of the present core.

Vityazev (1973) and Shaw (1978) both developed models in which iron separates out as a discrete layer as in Elsasser's model. However, in their models the layer as a whole falls to the centre of the Earth and not as a series of drops (Figure 1.3). In both their models, the time of core formation is estimated to be several hundred Ma. There is a difference however between their two models as to where in the Earth the energy of the falling iron is released. In Shaw's model, the core receives about one half of the energy, whilst in Vitazev's model, the mantle receives more than 80%. In Elsasser's model, the energy release is largest during the early stages of separation and then rapidly decreases. It must not be forgotten, however, that these models of core formation are highly oversimplified. This question is discussed further at the end of section 3.4.2.

A number of workers have concluded from studies of Pb isotope ratios that the time of core formation was quite short. Ringwood (1960) had argued that Fe, descending during core formation, would take with it substantial amounts of Pb but not U, i.e., core formation would change the Pb/U ratio in the upper crust and mantle. Hence the age of the Earth (~4550 Ma) as given by Pb/U geochronology, refers to the time of core formation which must have taken place very soon after the accretion of the Earth itself. If

Model#15   Model#30

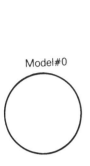
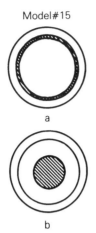
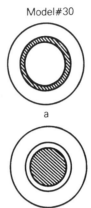
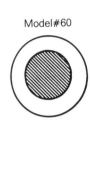

Model#0   Model#60

a   a

b   b

Figure 1.3 Earth models at various stages of core separation according to (a) the sinking layer model, (b) the Elsasser model. Model 15 has one half of the total mass differentiated. Model 30 shows differentiation to a depth of one half the Earth's radius (after Shaw, 1978).

Ringwood's argument is correct, then, during core formation the Pb/U ratio of the metal phase would be higher than that in silicates. This was confirmed by experimental work carried out by Oversby and Ringwood (1971). A relationship exists between the time interval between the accretion of the Earth and the formation of the core and the ratio of the Pb concentration in the metal to that in the silicate phase. This leads to the result that the Earth's core formed either during accretion or very soon after. Vollmer (1977) repeated their calculations using the revised decay constants of U, and obtained slightly different results. His models give no details of the process of core formation, but they also imply that core formation was fairly rapid, with a maximum time of 100 Ma between accretion of the Earth and the completion of core formation. The same conclusion has been reached by other authors, although allowing in some cases for the possibility of a slow, continuing growth of the core following a rapid initial growth. Allègre *et al.* (1982) estimated that 85% of the core would have formed early in the Earth's history (during the first 50–200 Ma), the remaining 15% over the rest of geologic time.

1.6 HEAT SOURCES IN THE EARLY EARTH

If the Earth accreted cold, what caused it to heat up sufficiently to melt at least the OC? We know from seismology (Chapter 2) that the OC is now liquid and probably has been so far at least 3500 Ma, since, as already mentioned, motions in a liquid core are needed to maintain the Earth's magnetic field. There is no shortage of possible candidates for a heat source, but it is not easy to quantify the magnitude of their effect in every case. In the first place the temperature within the aggregating Earth will rise by a certain amount (possibly by several hundred degrees) as a result of adiabatic compression as the proto-Earth grew.

Consider now possible heating that could arise during the accretion process. The importance of this as a source of heat depends on the details of the accretion process, which are but imperfectly known. The kinetic energy of the aggregating particles is either converted into internal energy or radiated away. How this division of energy is made depends on the temperature at the surface of the aggregating Earth and on the transparency of the surrounding atmosphere to radiation, neither of which are known with any certainty. Even more important is the duration of the accretion process – rapid accretion is essential to produce temperatures sufficiently high to melt the OC.

Another result of the accretion process that may be far more important, is the effect of impacts of falling bodies. Safronov (1978) was one of the first to stress the importance of impacts. He believed that relatively large bodies with diameters of several hundred kilometres were involved in the formation of the planets. If the impacting bodies were small, the heat generated by impact would quickly radiate into space with little heating of the Earth. The larger the impacting body, the greater the depth at which its energy would be released, and hence the greater the amount of this energy that would be trapped inside the Earth. In addition, larger bodies would cause deeper craters, leading to greater mixing of material on impact – heat transfer by mixing of material is far more efficient than heat transfer by thermal conduction. Safronov estimated that temperatures of the order of 1500 K would be generated at a depth of about 500 km in the mantle. His calculations have been refined by Kaula (1980) and Davies (1985). It seems that temperatures would never have risen much above melting since the viscosity of molten material would be low enough to remove by convection much of the heat generated by impacts. An important point is that a significant amount of impact energy can be retained in a slowly growing planet so that rapid formation is not required. A terrestrial planet would probably reach melting temperatures in its outer layers at a radius of between 2000 and 3000 km. The question of giant impacts in the late stages of the formation of the solar system has already been discussed in sections 1.2 to 1.4. It is generally agreed that, if giant impacts did occur, they would have caused global melting.

Finally if the Earth formed by accretion from homogeneous material and later differentiated into crust, mantle and core, the formation of the core would release a large amount of gravitational energy as the high density Ni−Fe sank to the centre of the Earth. Estimates of the energy release due to core formation indicate a temperature increase of about 1500−2000 K (Tozer, 1965; Flaser and Birch, 1973).

Consider now the decay of radioactive isotopes. In the first place it is necessary to distinguish between long-lived and short-lived radioactive isotopes. The possible importance of short-lived radioactive isotopes depends on the time interval Δt between the formation of the elements and the aggregation of the Earth. Δt must be short compared with the half-lives of the isotopes if they are to be a serious heat source in the early Earth. Possible short-lived isotopes are ^{236}U, ^{146}Sm, ^{244}Pu and ^{247}Cm all of which have half-lives sufficiently long to have contributed to heating up the Earth for some tens of millions of years after its formation. Three even shorter lived radionuclides ^{26}Al, ^{36}Cl and ^{60}Fe could have been a significant source for about 5–15 Ma after nucleosynthesis. The most important of all these short-lived radioactive isotopes is ^{26}Al because of its relatively large abundance in the Earth. ^{26}Al decays to ^{26}Mg with a half-life of 0.74 Ma and would remain a significant heat source for a few Ma. Anomalous values of $^{26}Mg/^{24}Mg$ have been found by Gray and Compston (1974) and Typhoon Lee *et al.* (1976) in chondrites in the Allende meteorite. The authors concluded that the most plausible cause of the ^{26}Mg anomaly is the decay *in situ* of ^{26}Al providing definite evidence of ^{26}Al in the early solar system. However, the source of the ^{26}Al which decayed to give the excess ^{26}Mg in the Allende meteorite is not clear − one suggestion is the explosion of a supernova beside the forming solar system. Large amounts of ^{26}Al have also now been found in the interstellar medium by HEO3 (the third High Energy Observatory). However the proposed excess ^{26}Al in Allende is too large to be the result of the averaged interstellar concentration observed so that, if real, some source of ^{26}Al enhancement in the collapsing solar nebula is necessary. Large excesses of radiogenic ^{26}Mg arising from the *in situ* decay of ^{26}Al have now been found in ten carbonaceous and ordinary chondritic meteorites. However, until recently, all observations of ^{26}Mg were from Al-rich minerals in refractory, Ca−Al rich, inclusions, making it difficult to assess the distribution of ^{26}Al in the solar system. Recently Hutcheon and Hutchison (1989) have found radiogenic ^{26}Mg in non-refractory meteorite material. They concluded that planetary accretion and differentiation must have begun on a timescale comparable to the half-life of ^{26}Al, and that, even if widespread melting did not occur, ^{26}Al heating played an important role in the thermal history of small planetary bodies.

Long-lived radioactive isotopes that are continuing to contribute to the heat budget of the Earth are ^{238}U, ^{235}U, ^{232}Th and ^{40}K, all of which have half-lives comparable to the age of the Earth. However because of their long half-lives, their contribution to the proto-Earth is small. After 100 Ma, the temperature increase would be about 150 K even assuming no heat loss. Thus, though important, the decay of long-lived radioactive isotopes is insufficient to provide the heat required to produce a liquid OC.

A number of other possible heat sources have been suggested, most of which are difficult to assess or appear to be inadequate, e.g., the dissipation of part of the Earth's rotational energy as it slows down through tidal interaction with the Moon. Another suggestion is the blanketing effect of a primordial dense atmosphere surrounding the proto-Earth (Hayashi *et al.*, 1979). A modification of this has been invoked by Matsui and Abe (1986) – a dense steam atmosphere resulting from the ejection from the Earth of large amounts of fine dust and the release of volatile gases such as H_2O and CO_2 following the impact of sizeable planetesimals. This would lead to a 'greenhouse effect', strong enough to trap the impact generated heat and melt the Earth. It is difficult to quantify the effect of such models, because of the unknown boundary conditions and composition of the Earth's proto-atmosphere.

2

The physical properties of the Earth

2.1 INTRODUCTION

The major source of our information about the Earth's interior comes from seismology – the study of earthquakes. The Earth is continually undergoing deformation due to stresses that build up within it. If the stresses continue for a long time, fracture may take place resulting in an earthquake with a sudden release of energy. Part of this energy is in the form of elastic waves which travel through the Earth carrying with them information about the structures through which they have passed. This information is recorded at the large number of seismological stations distributed around the world.

There are two basic types of elastic waves; body waves that propagate within a body of rock, and surface waves whose motion is restricted to near the surface. It is with body waves that we will be mainly concerned. The faster of the body waves is called the primary or P wave and is a longitudinal wave i.e., as it spreads out it alternately pushes (compresses) and pulls (dilates) the rock (Figure 2.1a). Thus, like sound waves, P waves can travel through both solid and liquid material. The slower body wave is called the secondary or S wave and, as it propagates, it shears the rock sideways at right angles to the direction of travel (Figure 2.1b). Like light waves, S waves are tranverse and hence are polarized.

2.2 TRAVEL-TIME AND VELOCITY-DEPTH CURVES

Knowing the speed of sound in sea water, echo sounders on ships can be used to map the depths of the ocean bottom. The same method can, in principle, be applied to determine structures deep within the Earth using a large earthquake as an energetic source and noting the time that it takes for seismic waves to be reflected back to the Earth's surface. However for this methods to be successful, it is first necessary to determine the velocity of seismic waves as a function of depth – in the case of the echo sounder, the velocity of sound in sea-water is already known.

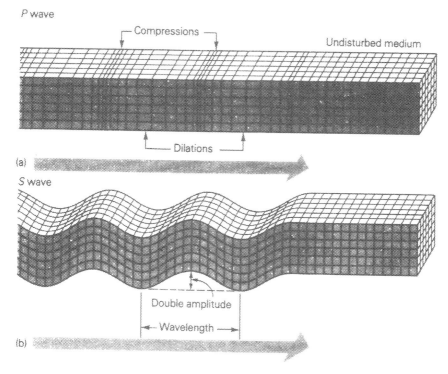

Figure 2.1 The two types of body waves (a) *P* waves, (b) *S* waves (after Bolt, 1976).

If the velocity of seismic waves throughout the Earth were known, then it would be possible to predict when a wave would arrive at any seismic observatory merely by dividing the distance from the source by the speed of the wave. This is an example of what is called a direct problem. In practice the situation is just the reverse. We can obtain measurements of the travel-times of earthquake waves at a finite number of stations and from these we have to determine both the velocity of waves through the Earth and its structure. This is an example of what is called an inverse problem and is the case in many geophysical situations. Such inverse problems do not have a unique solution. This is because we have only a finite number of observations which limit the resolution that can be obtained. In addition there will always be some errors in the observations themselves. Thus information about the deep structure of the Earth can only yield averaged or smoothed values and not an exact value at a particular depth.

It can be shown that the velocities of *P* and *S* waves are given by

$$V_p = \sqrt{\dfrac{k + \dfrac{4}{3}\mu}{\rho}} \qquad (2.1)$$

and
$$V_s = \sqrt{\dfrac{\mu}{\rho}} \qquad (2.2)$$

where ρ is the density, k the bulk modulus or (adiabatic) incompressibility and μ the rigidity. Thus V_p and V_s depend only on the elastic parameters and density of the medium through which they have passed. It can be seen that if the rigidity μ is zero, V_s is zero i.e., shear waves cannot be transmitted through a liquid. S waves have never been observed to pass through the outer core (OC) which is one of the main reason for believing that it is liquid. On the other hand, although S waves have not been positively identified as passing through the inner core (IC) we shall see later that there are good reasons for thinking that the IC is solid.

When an elastic wave meets a sharp boundary between two media of different properties, part of it will be reflected and part refracted and laws of reflection and refraction, analogous to those of light waves, apply. The case of elastic waves, however, is more complicated since waves of both P and S type may be reflected and

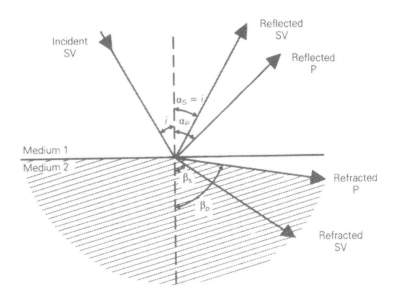

Figure 2.2 Reflected and refracted rays from an SV ray incident on a plane boundary between two solids (after Jacobs, 1974).

refracted. This could be expected since at the point of incidence the rock at the interface is being both compressed and sheared. The angles of reflection and refraction are related to the angle of incidence by the wave velocities V_p and V_s in the two media. Figure 2.2 shows the reflected and refracted rays from a vertically polarized S wave (SV) incident on a plane boundary between two solids. The relationships between the angles and velocities shown in the Figure are:

$$\frac{\sin i}{V_{S_1}} = \frac{\sin \alpha_S}{V_{S_1}} = \frac{\sin \alpha_P}{V_{P_1}} = \frac{\sin \beta_S}{V_{S_2}} = \frac{\sin \beta_P}{V_{P_2}} \tag{2.3}$$

There are three major discontinuities in the Earth – one just below the surface separating the crust from the mantle, another at a depth of about 2900 km separating the mantle from the core, and the third at a depth of about 5150 km dividing the core into an inner and an outer part. Figure 2.3 illustrates some of the many possible reflections and refractions at these boundaries and the terminology used to designate some of the different phases. Engdahl *et al.* in 1970 detected the phase PKiKP at the seismograph array LASA in Montana following nuclear explosions at the Nevada test site. This phase was reflected back at near vertical incidence from the boundary of the IC indicating that it has a sharp surface.

The times at which seismic waves from the same earthquake arrive at different observatories can be recorded so that it is possible to construct travel-time curves, i.e., plots of arrival times against distance from the source. Figure 2.4 illustrates travel time curves for P, S and surface waves where distance Δ is measured in degrees along the surface of the Earth. The curve for the first arrival of surface waves is a straight line while the curves for P and S waves have a downward concavity indicating that the velocity over most regions of the Earth increases with depth.

The travel times depend upon how the velocities of the different seismic waves change as they pass through materials with varying elastic properties (equations (2.1) and (2.2)). From the travel-time curves it is possible to obtain velocity-depth curves (Figure 2.5). The details of this 'inversion' will not be given here, but may be found in any standard text on seismology (e.g., Bullen, 1963). Velocity-depth curves provide the basic information for estimating the physical properties of the Earth's interior (section 2.4).

The velocity of seismic waves in general increases with depth. However, if there is a region in the Earth in which the velocity decreases with depth, the inversion method breaks down. In this case there will be a range of distances from the source in which

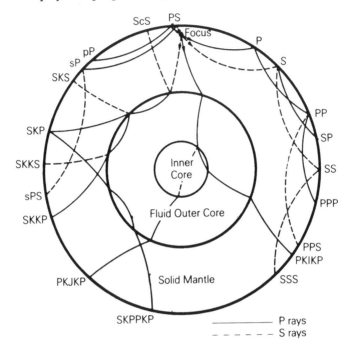

Figure 2.3 Representative seismic waves through the Earth and the terminology employed to designate the different phases. A *P* wave reflected from the Earth's surface can give rise to both a *P* wave and an *S* wave (called *PP* and *PS* waves respectively). Likewise an *S* wave reflected from the surface can give both *P* and *S* waves (called *SP* and *SS* waves respectively). The letters *c* and *i* denote reflections from the outer and inner core boundaries, respectively, while the letters *K* and *I* indicate *P* waves through the outer and inner core. Thus ScS is an *S* wave that has travelled down to the core boundary and been reflected as an *S* wave. Similarly SKP is an *S* wave that has been refracted into the outer core (necessarily as a *P* wave) and refracted back into the mantle as a *P* wave. *S* waves have never been observed in the outer core which is the main reason for believing it to be fluid. On the other hand, it is now believed that the inner core is solid and the symbol *J* has been proposed for paths of *S* waves (if they exist) in the inner core. (After Bullen, 1954.)

direct arrivals of *P* and *S* waves are not observed – a 'shadow zone'. There is an abrupt discontinuous drop in the velocity of *P* waves across the boundary between the mantle and core leading to such a shadow zone (Figure 2.6). The Danish seismologist Inge Lehmann (who celebrated her 100th birthday in 1988) suggested in 1935 that certain of the waves recorded in the shadow zone had passed through an inner core (IC) in which the *P* velocity is significantly

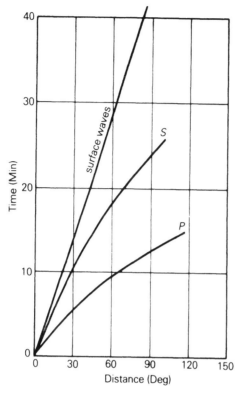

Figure 2.4 Travel-time curves for P, S and surface waves. The curves for other phases are more complicated.

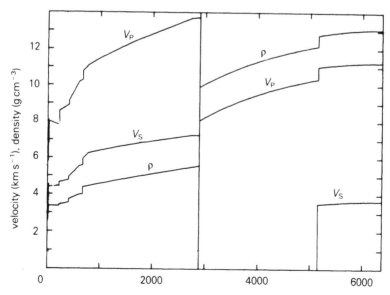

Figure 2.5 Seismic velocities of P and S waves and density ρ as a function of depth (after Dziewsonki and Anderson, 1981).

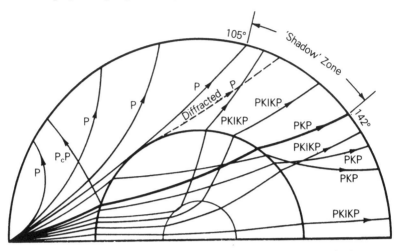

Figure 2.6 *P*, *PcP*, *PKP* and *PKIKP* waves, showing the shadow zone (after Bullen, 1954).

greater than that in the outer core (OC). Later work has confirmed this hypothesis and shown that the IC is solid.

2.3 FREE OSCILLATIONS OF THE EARTH

When any point of a deformable body like the Earth is set in motion, it may be regarded, either as a medium transmitting waves or as a vibrating system. Seismic wave theory, which we have discussed in section 2.2, considers the motions of the Earth following an earthquake as travelling disturbances which affect only a relatively small part of the Earth's volume at any given time. Oscillation theory, on the other hand, regards the motions as normal modes of oscillation, the principal ones of which affect a relatively large fraction of the volume of the Earth at any instant. The wave theory approach is more appropriate for short periods ($\lesssim$ 3 min), the oscillation approach for longer periods. The relatively late application of oscillation theory to the Earth is due to the fact that only recently have long period seismographs been developed.

The Earth literally rings like a bell following a great earthquake, the vibrations lasting several days. The tones are too low for the human ear to detect but can be observed on modern, sensitive seismographs. The 'football' mode of vibration, to be described later, has a period of ~54 min, corresponding to E flat in the twentieth octave below middle C! The vibrations of an elastic sphere have been studied by mathematicians since the 18th Century and in 1911 Love predicted that a steel sphere the size of the Earth would have a fundamental period of ~1 hour with overtones of lesser periods. In 1954, Benioff claimed to have detected an oscillation with a period of 57 min on a seismogram following the Kamchatka earthquake of November 4th, 1952. This was confirmed later on records from North and South America, Europe and Japan after the great Chilean earthquake of May 22nd, 1960. Earth vibrations following the Good Friday earthquake in Alaska (March 28th, 1964) were seen on long period seismographs at over 100 stations around the world. A perfect spherical bell would produce a pure tone, but the rotation of the Earth distorts its shape and creates additional tones. The various modes differ in their sensitivity to 'imperfections' at different locations within the Earth. Thus analysis of the free oscillations of the Earth can help locate anomalous structures in the Earth.

From an analysis of the normal modes of the Good Friday earthquake, Dziewonski and Gilbert (1971) concluded that the IC of the Earth must be solid. With the aid of modern computers, calculations have now been made of the expected frequencies and amplitudes of a number of Earth models, taking into account detailed structure, gravitational forces and rotation. Although such calculations are of value in comparing an Earth model with observations, we have not addressed the real problem, *viz.*, starting with the observed oscillations, to try and determine the structure and elastic properties of the Earth. This is another example of an inverse problem with all the inherent difficulties of uniqueness and resolution.

There are two separate types of free oscillations of an elastic sphere – spheroidal (*S*) and torsional (*T*) (sometimes called toroidal). Torsional oscillations have tangential but no radial displacements. Since the dilatation is zero, torsional oscillations cause no disturbances in density and hence no changes in the gravitational field. Thus any instruments designed to measure small fluctuations in gravity cannot record torsional oscillations. In spheroidal vibrations, the displacements have, in general, both radial and horizontal components. Like *P* and *S* waves, spheroidal and torsional oscillations generally occur together, i.e., they are coupled and their periods overlap.

Consider first for simplicity the vibrations of a string (Figure 2.7). The nodes are points that do not move. There is a similar series of nodal lines on the surface of a vibrating sphere, more or less corresponding to lines of latitude and longitude (Figure 2.8). A seismograph on the surface of the Earth on a nodal line of a

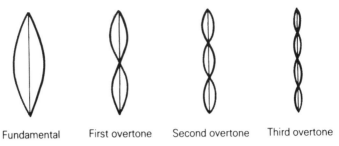

| Fundamental | First overtone | Second overtone | Third overtone |

Figure 2.7 The figure shows the modes of vibration of a string. The string moves from the position shown in black to the position shown in grey, and back again. The nodes are the points that do not move (after Bolt, 1982).

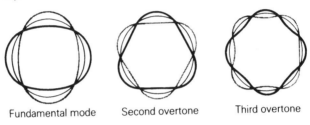

| Fundamental mode | Second overtone | Third overtone |

Figure 2.8 Showing the surface view of spheroidal vibrations of a sphere. Points at which the black and grey lines intersect are nodal lines in the surface of the sphere (after Bolt, 1982).

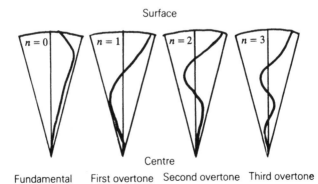

Surface

$n = 0$ $n = 1$ $n = 2$ $n = 3$

Centre

| Fundamental | First overtone | Second overtone | Third overtone |

Figure 2.9 Depicting the interior view of the radial modes of vibration of a sphere. The wavy line indicates the variation in displacement with depth in a homogeneous sphere. In the Earth, the displacements are complicated by the liquid core (after Bolt, 1982).

particular mode, would not record that mode. In addition there will, in general, be spherical surfaces inside the sphere where the various modes of vibration have no displacement (Figure 2.9).

Spheroidal and torsional modes are described by the symbols $_nS_1$ and $_nT_1$. The first suffix n denotes the number of internal nodal surfaces, not counting the one at the Earth's centre. The second suffix 1 is the number of surface nodal lines for spheroidal oscillations and one more for torsional oscillations, i.e., it is the number of separate sectors on the surface. The simplest (fundamental) torsional mode is $_0T_2$. There is no displacement at the equator (there is only one nodal line), and the two hemispheres oscillate in antiphase (Figure 2.10). Figure 2.11 shows the first radial overtone $_1T_2$. There is still only one surface nodal line ($1 = 2$), but there is also now one interior (radial) surface ($n = 1$) across which the motion changes direction. The fundamental spheroidal oscillation $_0S_0$ is an alternating compression and expansion of the whole sphere. The next higher mode is $_0S_2$. There are two surface nodal lines corresponding to lines of latitude in the north and south hemispheres (Figure 2.12). This is known as the 'football' mode – as the sphere oscillates it is distorted alternately into an oblate and prolate spheroid. As with torsional oscillations there are also internal nodal surfaces – Figure 2.13 shows the first radial overtone $_1S_3$.

There is also fine structure in the frequency spectrum of the Earth's vibrations. The single peak near a period of 53.8 min for $_0S_2$ is split into at least three peaks (Figure 2.14). The difference in frequency between the longest and shortest vibrations in this multiplet is a few per cent. Pekeris *et al.* (1961) suggested that this might be the result of the rotation of the Earth splitting any mode having a longitude dependence into a number of frequencies. They coined the term 'terrestrial spectroscopy' for the study of free oscillation spectra – the observed splitting of certain lines due to the Earth's rotation being a mechanical analogue of the Zeeman splitting of lines of atomic spectra by a magnetic field. In addition, any deviation from perfect elasticity and sphericity will split the spectral lines. The Earth is not a perfect sphere, and approximately 70% of its surface is covered by oceans – mostly in the southern hemisphere. Both its ellipticity and the relative positions of the continents and oceans affect the frequency spectrum of the Earth's free oscillations. The very long period modes ($_0T_2$, $_0S_2$ . . .) that involve the vibration of the bulk of the Earth are mostly affected by rotation. For shorter periods ($_0S_{10}$, $_0T_{10}$ . . .), the rotation is confined more and more to the upper parts of the Earth, and the effects of ellipticity and tectonic differences become more important.

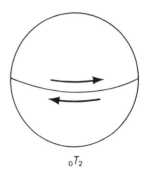

$_0T_2$

Figure 2.10 Representation of the mode $_0T_2$ (after Bolt, 1982).

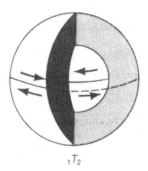

$_1T_2$

Figure 2.11 Representation of the mode, $_1T_2$ (after Bolt, 1982).

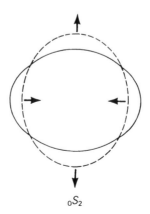

$_0S_2$

Figure 2.12 Representation of the mode $_0S_2$ (after Bolt, 1982).

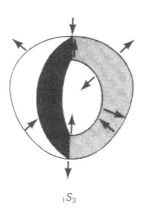

$_1S_3$

Figure 2.13 Representation of the mode, $_1S_3$ (after Bolt, 1982).

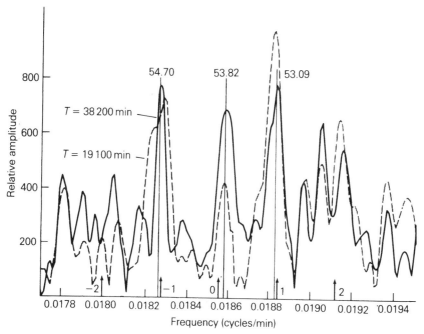

Figure 2.14 The fine frequency structure for $_0S_2$ obtained by a Fourier analysis of two lengths of record, 19 100 and 38 200 minutes from the Isabella strain meter after the Chilean earthquake of May 22, 1960. Sample interval is 2 minutes. Theoretical splitting from the rotation of the Earth is marked on the base line (after Bolt, 1982).

Modern seismographs have detected certain modes for weeks following a large earthquake. Ultimately however the energy of the vibrations is slowly dissipated into heat by non-elastic processes in the interior of the Earth – and this can also tell us something about the structure and constitution of the Earth. Departures from perfect elasticity in the Earth can be expressed in terms of the reciprocal of the dimensionless parameter Q (Knopoff, 1964), defined by the equation

$$\frac{2\pi}{Q} = \frac{\Delta E}{E} \tag{2.4}$$

where ΔE is the energy dissipated per cycle and E is the peak energy stored. The mechanical Q defined above is analogous to the Q of an oscillating electrical circuit. For estimating Q values in different regions of the Earth, use has been made of both body waves and free oscillations.

Before we leave this section it must be pointed out that the 'average' Earth to which the average periods of free oscillation modes relate is not necessarily the same as an Earth model obtained from the average travel times of seismic body waves. This is because earthquake epicentres and recording stations are not uniformly spaced over the Earth's surface, but are irregularly distributed. It is also worthwhile stressing again that since the variations in the physical properties of the Earth are in general continuous, and the amount of data that we can collect is of necessity finite, there is no unique solution to the problem of determining the constitution of the Earth. The problem is further compounded by the fact that the data themselves are not free from error. However it is possible to place bounds on possible solutions and certain properties of the Earth are known to a high degree of accuracy.

2.4 VARIATION OF DENSITY AND OTHER PHYSICAL PROPERTIES WITHIN THE EARTH

The density ρ will depend on the pressure p, temperature T and the chemical composition which we can denote by a number of parameters n_1 (i = 1, 2 . . .), i.e., we can write

$$\rho = \rho(p, T, n_i) \tag{2.5}$$

If m is the mass of material within a sphere of radius r, then, since the stress in the Earth's interior is essentially equivalent to a hydrostatic pressure

$$\frac{dp}{dr} = -g\rho \tag{2.6}$$

where g is the acceleration due to gravity (or force on unit mass) i.e.,

$$g = \frac{Gm}{r^2} \tag{2.7}$$

and G the constant of gravitation. The assumption of hydrostatic pressure would be a poor approximation for stresses in the Earth's crust, but is unlikely to be much in error in the mantle and core.

Consider a chemically homogeneous layer in the Earth, i.e., one in which there are no significant changes in phase or chemical composition and in which the temperature variation is adiabatic (i.e., there is no change in heat). Then it follows from equation (2.5) that

$$\frac{d\rho}{dr} = \frac{d\rho}{dp} \cdot \frac{dp}{dr} = \frac{-g\rho^2}{k} \tag{2.8}$$

where k is the incompressibility or bulk modulus defined by

$$\frac{1}{k} = \frac{1}{\rho}\frac{d\rho}{dp} \tag{2.9}$$

i.e. k is the ratio of the pressure change to the compression produced. From equations (2.1) and (2.2)

$$V_P^2 - \frac{4}{3}V_S^2 = \frac{k}{\rho} \tag{2.10}$$

so that the distribution of k/ρ denoted by ϕ throughout the Earth is known from the velocity-depth curves. Finally combining equations (2.7), (2.8) and (2.10) we have

$$\frac{d\rho}{dr} = \frac{-Gm\rho^2}{kr^2} = \frac{-Gm\rho}{r^2\phi} \tag{2.11}$$

Since $dm/dr = 4\pi\rho r^2$, a second order differential equation for $\rho = \rho(r)$ can be obtained by differentiating equation (2.11). This equation was first obtained by Williamson and Adams in 1923 and may be integrated numerically to obtain the density distribution in the Earth. Much of the early work was carried out by Bullen, a good account of which has been given by Bullen himself (1975). These earlier determinations of the density distribution only made use of the data contained in the velocity-depth curves. This was supplemented later by experimental empirical data which indicated an approximate linear relationship between V_p and ρ. However the largest source of additional information has come from analyses of the free oscillations of the Earth following large earthquakes. A comparison of the theoretical periods calculated for Earth models based on velocity depth curves showed reasonable agreement with the observed periods (within ~1%). However, more precise measurements later showed some discrepancies, e.g., none of the existing velocity-depth curves combined with Bullen's density distribution was compatible with the longer period free oscillations obtained from the Chilean earthquake of May 22, 1960 and the Alaskan earthquake of March 28, 1964.

A different approach to the inverse problem is the Monte Carlo method pioneered by Press (1968). This method uses random selection to generate large numbers of models in a computer, subjecting each model to a test against geophysical data (velocity-depth curves, free oscillation data, total mass and moment of inertia). Only those models whose properties fit the data within prescribed limits are accepted. Such a procedure has the advantage of finding models without any bias from preconceived or oversimplified ideas of Earth structure. Press tested more than one million models and was able to find a large number of successful ones. Some of the 'successful' models were rather 'spiky' and improbable – this is understandable since a simple random walk would automatically have a low probability. An alternative approach was adopted by Bullen and Haddon (1970). Their procedure was to start with models based only on the velocity-depth curves and then derive a sequence of models showing improved agreement with the free oscillation data, i.e., they would vary one or more parameters at each stage and try to satisfy the oscillation data.

A general discussion of the inverse problem in geophysics has been given in a series of papers by Backus and Gilbert, the details of which are beyond the scope of this book. For an account of the method and references, see Parker (1972). It must be stressed again that the inverse problem has no unique solution. Moreover the data will be subject to some error. We can improve on the error estimate of a given property deduced from the data, but only at the expense of the resolving power and *vice versa*. There is thus a trade-off between the ability to resolve detail and the accuracy with which this detail can be estimated. Free oscillation data are not adequate to resolve details of Earth structure having wave lengths of the order of 100–200 km. To resolve these features body wave data which are of higher resolving power must be used.

There has been a proliferation of Earth models, the details of which will not be described here, with the result that it is not easy for the reader to choose between them. There is also the added (and perhaps not appreciated) danger of adopting some properties from one model and other properties from another with the result that the combined model is not self-consistent. To try and overcome these difficulties, the International Union of Geodesy and Geophysics (IUGG) at their XV General Assembly in Moscow in 1971, set up a committee to advise on a standard earth model (SEM). After many meetings and preliminary Earth models, Dziewonski and Anderson were given the task of producing the final model. They presented a preliminary reference earth model (PREM) at the XVII General Assembly of the IUGG in Canberra in 1979. Their results were later published in 1981 (Table 2.1). It was appreciated that refinements of PREM are necessary, e.g., it is now possible to take account of damping. However the effects of attenuation are not well-determined and, although a definite SEM has not yet been produced, the SEM committee was finally dissolved at the XVIII General Assembly of the IUGG in Hamburg in 1983.

Once the density distribution has been obtained, the pressure distribution and the elastic constants can be computed. From equations (2.6) and (2.7) it follows that

$$\frac{dp}{dr} = \frac{-Gm\rho}{r^2} \tag{2.12}$$

Since the density is used only to determine the pressure gradient, the pressure distribution is insensitive to small changes in the density distribution. Equations (2.2) and (2.10) give μ and k directly. It can be shown that Poisson's ratio σ is given by

$$\sigma = \frac{3k - 2\mu}{6k + 2\mu} = \frac{V_P^2 - 2V_s^2}{a(V_P^2 - V_s^2)} \tag{2.13}$$

on using equations (2.1) and (2.2). Thus σ can be determined directly from the velocity-depth curves and is independent of the density ρ. Finally, the variation of g can be calculated from equation (2.7). It is interesting that its value does not differ by more than one per cent from $990 \, \text{cm s}^{-2}$ until a depth of over $2400 \, \text{km}$ is reached. It actually starts increasing with depth, reaching a value of $\sim 1068 \, \text{cm s}^{-2}$ at a depth of $2890 \, \text{km}$, before decreasing to zero at the Earth's centre.

Finally it must be pointed out that although seismology has enabled us to estimate the density and other physical properties within the Earth, it does not allow us to determine the chemical composition or the temperature of the Earth. These questions will be addressed in Chapters 3 and 4.

Table 2.1. Earth model PREM. Above 220 km the mantle is transversely isotropic; the parameters given are 'equivalent' isotropic moduli and velocities (after Dziewonski and Anderson, 1981).

Radius (km)	Depth (km)	Density (g cm⁻³)	V_p (km s⁻¹)	V_s (km s⁻¹)	φ (km² s⁻²)	κ (kbar)	μ (kbar)	σ	Pressure (kbar)	dk/dp	Gravity (cm s⁻²)
0.	6371.0	13.08848	11.26220	3.66780	108.90	14253	1761	0.4407	3638.524	2.3360	0.
100.0	6271.0	13.08630	11.26064	3.66670	108.88	14248	1759	0.4407	3636.131	2.3363	36.56
200.0	6171.0	13.07977	11.25593	3.66342	108.80	14231	1755	0.4408	3628.956	2.3365	73.11
300.0	6071.0	13.06888	11.24809	3.65794	108.68	14203	1749	0.4409	3617.011	2.3369	109.61
400.0	5971.0	13.05364	11.23712	3.65027	108.51	14164	1739	0.4410	3600.315	2.3375	146.04
500.0	5871.0	13.03404	11.22301	3.64041	108.29	14114	1727	0.4412	3578.894	2.3382	182.39
600.0	5771.0	13.01009	11.20576	3.62835	108.02	14053	1713	0.4414	3552.783	2.3391	218.62
700.0	5671.0	12.98178	11.18538	3.61411	107.70	13981	1696	0.4417	3522.024	2.3402	254.73
800.0	5571.0	12.94912	11.16186	3.59767	107.33	13898	1676	0.4420	3486.665	2.3414	290.68
900.0	5471.0	12.91211	11.13521	3.57905	106.91	13805	1654	0.4424	3446.764	2.3428	326.45
1000.0	5371.0	12.87073	11.10542	3.55823	106.45	13701	1630	0.4428	3402.383	2.3443	362.03
1100.0	5271.0	12.82501	11.07249	3.53522	105.94	13586	1603	0.4432	3353.596	2.3460	397.39
1200.0	5171.0	12.77493	11.03643	3.51002	105.38	13462	1574	0.4437	3300.480	2.3480	432.51
1221.5	5149.5	12.76360	11.02827	3.50432	105.25	13434	1567	0.4438	3288.513	2.3486	440.02
1221.5	5149.5	12.16634	10.35568	0.	107.24	13047	0	0.5000	3288.502	3.7545	440.03
1300.0	5071.0	12.12500	10.30971	0.	106.29	12888	0	0.5000	3245.423	3.6539	463.68
1400.0	4971.0	12.06924	10.24959	0.	105.05	12679	0	0.5000	3187.493	3.5478	494.13
1500.0	4871.0	12.00989	10.18743	0.	103.78	12464	0	0.5000	3126.159	3.4649	524.77
1600.0	4771.0	11.94682	10.12291	0.	102.47	12242	0	0.5000	3061.461	3.4017	555.48
1700.0	4671.0	11.87900	10.05572	0.	101.12	12013	0	0.5000	2993.457	3.3552	586.14
1800.0	4571.0	11.80900	9.98554	0.	99.71	11775	0	0.5000	2922.221	3.3230	616.69
1900.0	4471.0	11.73401	9.91206	0.	98.25	11529	0	0.5000	2847.838	3.3028	647.04
2000.0	4371.0	11.65478	9.83496	0.	96.73	11273	0	0.5000	2770.407	3.2927	677.15
2100.0	4271.0	11.57119	9.75393	0.	95.14	11009	0	0.5000	2690.035	3.2911	706.57

2200.0	4171.0	11.48311	9.66865	0.	93.48	10735	0	0.5000	2606.838	3.2966	736.45
2300.0	4071.0	11.39042	9.57881	0.	91.75	10451	0	0.5000	2520.942	3.3080	765.56
2400.0	3971.0	11.29298	9.48409	0.	89.95	10158	0	0.5000	2432.484	3.3242	794.25
2500.0	3871.0	11.19067	9.38418	0.	88.06	9855	0	0.5000	2341.603	3.3441	822.48
2600.0	3771.0	11.08335	9.27876	0.	86.10	9542	0	0.5000	2248.453	3.3670	850.23
2700.0	3671.0	10.97091	9.16752	0.	84.04	9220	0	0.5000	2153.189	3.3919	877.46
2800.0	3571.0	10.85321	9.05015	0.	81.91	8889	0	0.5000	2055.978	3.4180	904.14
2900.0	3471.0	10.73012	8.92632	0.	79.68	8550	0	0.5000	1956.991	3.4448	930.23
3000.0	3371.0	10.60152	8.79573	0.	77.36	8202	0	0.5000	1856.409	3.4714	955.70
3100.0	3271.0	10.46727	8.65805	0.	74.96	7846	0	0.5000	1754.418	3.4972	980.51
3200.0	3171.0	10.32726	8.51298	0.	72.47	7484	0	0.5000	1651.209	3.5215	1004.64
3300.0	3071.0	10.18134	8.36019	0.	69.89	7116	0	0.5000	1546.982	3.5437	1028.04
3400.0	2971.0	10.02940	8.19939	0.	67.23	6743	0	0.5000	1441.941	3.5629	1050.65
3480.0	2891.0	9.90349	8.06482	0.	65.04	6441	0	0.5000	1357.510	3.5769	1068.23
3480.0	2891.0	5.56645	13.71660	7.26466	117.78	6556	2938	0.3051	1357.509	1.6435	1068.23
3500.0	2871.0	5.55641	13.71168	7.26486	117.64	6537	2933	0.3049	1345.619	1.6434	1065.32
3600.0	2771.0	5.50642	13.68753	7.26575	116.96	6440	2907	0.3038	1287.067	1.6424	1052.04
3630.0	2741.0	5.49145	13.68041	7.26597	116.76	6412	2899	0.3035	1269.742	1.6420	1048.44
3630.0	2741.0	5.49145	13.68041	7.26597	116.76	6412	2899	0.3035	1269.741	3.3344	1048.44
3700.0	2671.0	5.45657	13.59597	7.23403	115.08	6279	2855	0.3026	1229.719	3.2957	1040.66
3800.0	2571.0	5.40681	13.47742	7.18892	112.73	6095	2794	0.3012	1173.465	3.2443	1030.95
3900.0	2471.0	5.35706	13.36074	7.14423	110.46	5917	2734	0.2998	1118.207	3.2029	1022.72
4000.0	2371.0	5.30724	13.24532	7.09974	108.23	5744	2675	0.2984	1063.864	3.1716	1015.80
4100.0	2271.0	5.25729	13.13055	7.05525	106.04	5575	2617	0.2971	1010.363	3.1503	1010.06
4200.0	2171.0	5.20713	13.01579	7.01053	103.88	5409	2559	0.2957	957.641	3.1393	1005.35
4300.0	2071.0	5.15669	12.90045	6.96538	101.73	5246	2502	0.2943	905.646	3.1383	1001.56
4400.0	1971.0	5.10590	12.78389	6.91957	99.59	5085	2445	0.2928	854.332	3.1472	998.59
4500.0	1871.0	5.05469	12.66550	6.87289	97.43	4925	2388	0.2913	803.660	3.1657	996.35
4600.0	1771.0	5.00299	12.54466	6.82512	95.26	4766	2331	0.2898	753.598	3.1935	994.74
4700.0	1671.0	4.95073	12.42075	6.77606	93.06	4607	2273	0.2881	704.119	3.2302	993.69

Table 2.1. Continued

Radius (km)	Depth (km)	Density (g cm⁻³)	V_p (km s⁻¹)	V_s (km s⁻¹)	ϕ (km² s⁻²)	κ (kbar)	μ (kbar)	σ	Pressure (kbar)	dk/dp	Gravity (cm s⁻²)
4800.0	1571.0	4.89783	12.29316	6.72548	90.81	4448	2215	0.2864	655.202	3.2750	993.14
4900.0	1471.0	4.84422	12.16126	6.67317	88.52	4288	2157	0.2846	606.830	3.3276	993.01
5000.0	1371.0	4.78983	12.02445	6.61891	86.17	4128	2098	0.2826	558.991	3.3871	993.26
5100.0	1271.0	4.73460	11.88209	6.56250	83.76	3966	2039	0.2805	511.676	3.4527	993.83
5200.0	1171.0	4.67844	11.73357	6.50370	81.28	3803	1979	0.2783	464.882	3.5236	994.67
5300.0	1071.0	4.62129	11.57828	6.44232	78.72	3638	1918	0.2758	418.606	3.5989	995.73
5400.0	971.0	4.56307	11.41560	6.37813	76.08	3471	1856	0.2731	372.852	3.6775	996.98
5500.0	871.0	4.50372	11.24490	6.31091	73.34	3303	1794	0.2701	327.623	3.7582	998.36
5600.0	771.0	4.44317	11.06557	6.24046	70.52	3133	1730	0.2668	282.928	3.8403	999.85
5600.0	771.0	4.44316	11.06556	6.24046	70.52	3133	1730	0.2668	282.927	2.9819	999.85
5650.0	721.0	4.41241	10.91005	6.09418	69.51	3067	1639	0.2732	260.783	3.0086	1000.63
5701.0	670.0	4.38071	10.75131	5.94508	68.47	2999	1548	0.2798	238.342	3.0358	1001.43
5701.0	670.0	3.99214	10.26622	5.57020	64.03	2556	1239	0.2914	238.334	2.4000	1001.43
5736.0	635.0	3.98399	10.21203	5.54311	63.32	2523	1224	0.2911	224.364	2.3868	1000.88
5771.0	600.0	3.97584	10.15782	5.51602	62.61	2489	1210	0.2909	210.426	2.3734	1000.38
5771.0	600.0	3.97584	10.15782	5.51600	62.61	2489	1210	0.2909	210.425	8.0910	1000.38
5821.0	550.0	3.91282	9.90185	5.37014	59.60	2332	1128	0.2917	190.703	7.8833	999.65
5871.0	500.0	3.84980	9.64588	5.22428	56.65	2181	1051	0.2924	171.311	7.6761	998.83
5921.0	450.0	3.78678	9.38990	5.07842	53.78	2037	977	0.2933	152.251	7.4695	997.90
5971.0	400.0	3.72378	9.13397	4.93259	50.99	1899	906	0.2942	133.527	7.2633	996.86
5971.0	400.0	3.54325	8.90522	4.76989	48.97	1735	806	0.2988	133.520	3.3718	996.86

6016.0	355.0	3.51639	8.81867	4.73840	47.83	1682	790	0.2971	117.702	3.3369	995.22
6061.0	310.0	3.48951	8.73209	4.70690	46.71	1630	773	0.2952	102.027	3.3017	993.61
6106.0	265.0	3.46264	8.64552	4.67540	45.60	1579	757	0.2933	86.497	3.2662	992.03
6151.0	220.0	3.43578	8.55896	4.64391	44.50	1529	741	0.2914	71.115	3.2305	990.48
6151.0	220.0	3.35950	7.98970	4.41885	37.80	1270	656	0.2797	71.108	-0.7364	990.48
6186.0	185.0	3.36330	8.01180	4.43108	38.01	1278	660	0.2797	59.466	-0.7200	989.11
6221.0	150.0	3.36710	8.03370	4.44361	38.21	1287	665	0.2796	47.824	-0.7035	987.83
6256.0	115.0	3.37091	8.05540	4.45643	38.41	1295	669	0.2795	36.183	-0.6868	986.64
6291.0	80.0	3.37471	8.07688	4.46953	38.60	1303	674	0.2793	24.546	-0.6700	985.53
6291.0	80.0	3.37471	8.07689	4.46954	38.60	1303	674	0.2793	24.539	-0.6700	985.53
6311.0	60.0	3.37688	8.08907	4.47715	38.71	1307	677	0.2793	17.891	-0.6603	984.93
6331.0	40.0	3.37906	8.10119	4.48486	38.81	1311	680	0.2792	11.239	-0.6505	984.37
6346.6	24.4	3.38076	8.11061	4.49094	38.89	1315	682	0.2790	6.043	-0.6428	983.94
6346.6	24.4	2.90000	6.80000	3.90000	25.96	753	441	0.2789	6.040	-0.0000	983.94
6356.0	15.0	2.90000	6.80000	3.90000	25.96	753	441	0.2549	3.370	0.0000	983.32
6356.0	15.0	2.60000	5.80000	3.20000	19.99	520	266	0.2549	3.364	0.0000	983.31
6368.0	3.0	2.60000	5.80000	3.20000	19.99	520	266	0.2812	0.303	-0.0000	982.22
6368.0	3.0	1.02000	1.45000	0.	2.10	21	0	0.2812	0.299	-0.0000	982.22
6371.0	0.	1.02000	1.45000	0.	2.10	21	0	0.5000	-0.000	0.000	981.56

2.5 THE CORE-MANTLE BOUNDARY (CMB)

The most notable feature of the velocity-depth curves is the discontinuity at a depth of ~2900 km marking the boundary between
the mantle and the core. There is a discontinuous drop in the
velocity of P waves across the boundary and below it S waves have
never been observed. This is the main reason for believing that
the OC is liquid. There is a smaller discontinuity just below the
Earth's surface marking the boundary between the crust and mantle
(the Mohorovičić discontinuity), a low velocity layer varying in
depth between about 50 and 250 km and regions of higher velocity
gradients in the upper mantle near depths of 400 and 670 km which
are usually interpreted as phase transitions. These lesser discontinuities will not be discussed in detail in this book which is mainly
concerned with the Earth's core. We will refer to them again in
section 6.6 when we discuss convection in the mantle.

The CMB separates two dynamic systems with very different
compositions and material properties. There has been much argument about the structure of the bottom 200–300 km of the mantle
(called by Bullen the D″ layer). Constraints on the properties of this
transition zone come from a number of geophysical disciplines –
seismology, mineral physics, geomagnetism and geodynamics.
Seismology has provided the longest record of observations, but
there is still no agreement on their interpretation. Some seismologists maintain that the velocity of both P and S waves decreases as
the CMB is approached, others that the velocities remain almost
constant and others that the velocities continue to increase slightly.
The main unresolved problem is whether this transition zone is a
thermal or a chemical boundary layer. The question will be considered later in Chapter 6 after the thermal regime of the Earth and
its magnetic field have been discussed. The structure of the CMB is
a very active field of research in geophysics today. It influences the
mode of mantle convection, plume formation, the generation of the
Earth's magnetic field, changes in the length of the day and the
chemical evolution of the Earth. The importance of this region has
now been recognized by the geophysical community and in 1987
the IUGG set up a committee SEDI (study of the Earth's deep
interior) to study these questions. National committees have also
been established.

Before we leave this section, let us consider the question of the
inner core boundary (ICB). We have already mentioned that we
now believe the IC to be solid. Bullen in 1953 put forward a very
simple and convincing argument based on the velocity-depth curves
that this was indeed the case. The seismic P velocity V_p increases

suddenly at the ICB from about 10.3 to 11.0 km s^{-1}, i.e., about 6%, so that V_p^2 jumps by about 12%. In a liquid $V_s = 0$, so that from equation (2.10).

$$V_p^2 = \frac{k}{\rho} \qquad (2.14)$$

Since it is almost impossible for the density ρ to decrease with increasing depth, this means that the increase in V_p^2 implies a sudden increase of at least 12% in the incompressibility k. This contradicts separate physical evidence that k changes very little at the extreme pressures in the core for all likely core materials. However, if the IC is solid and hence has finite rigidity μ and is thus capable of transmitting S waves, then equation (2.14) is replaced by equation (2.10) and the jump in V_p^2 can be accounted for by the term $4/3 V_s^2$.

2.6 SEISMIC TOMOGRAPHY

Three-dimensional analyses of Earth structure are now possible as a result of the development of seismometers with proper frequency response and dynamic range which record data in digital form that can be fed into a computer. High-speed computers can now handle the immense amount of data and complex calculations necessary in three-dimensional problems. More than 2 million P wave travel time data have been inverted to obtain three-dimensional velocity pertubations in the Earth's mantle (Inoue *et al.*, 1990). By analogy with X-ray tomography which can reveal images of specific plane sections of the body, the technique of mapping the 3-dimensional structure of the Earth has been called seismic tomography.

If the internal properties of the Earth were spherically symmetric, our planet would be tectonically dead. That this is not so is evidenced by earthquakes, volcanoes, and, on a longer time scale, by mountain building and sea-floor spreading. Such dynamic behaviour must be driven by lateral differences in temperature and density. Seismic tomography has now made 3-D maps of the Earth's interior from the bottom of the crust to the IC.

Seismic tomography should give better information about shallow rather than deep regions, since earthquakes, the source of the energy, are confined to shallow depths. Most deep earthquakes occur in subducted plates where there are sharp lateral temperature, and hence seismic velocity, changes. These effects could dominate the true signal from the deep interior of the Earth with the danger that shallow structures could be mapped into deep ones. Morelli and Dziewonski (1987) have managed to avoid this problem in their

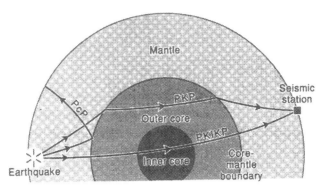

Figure 2.15 PcP, PKP and PKIKP waves.

mapping of the CMB by using the travel times of seismic waves that either travel through the liquid core (PKP) or have been reflected from the CMB (PcP) (Figure 2.15). Reflected waves will arrive early if there is a bump on the CMB, because it has less far to travel. A wave travelling through the same bump to the opposite side of the world will arrive late because waves travel slower in the core than in the lower mantle. The effect of a bump on the CMB is thus opposite for the two waves, but the effect of any heterogeneity in the mantle should be the same. However, interpretation of the results is not straightforward, e.g., a PKP wave may be slowed by a bump on the CMB, by its passage through a region of the mantle that is hotter and hence less rigid, or through a region of the mantle or core whose chemical composition is slightly different from that of its surroundings. Nevertheless maps of the CMB based on these two separate data sets showed good agreement and are thus likely to reflect true topography of the CMB and not any mantle heterogeneity. Morelli and Dziewonski found the relief of the CMB to be ±6 km with no detectable evidence of lateral heterogeneity in the liquid OC.

Undulations of the CMB may not account for all of the seismic variability seen there. 'Debris' may have collected at the CMB from above and below much as it has done at the Earth's surface. Any seismic wave from an earthquake passing through the CMB will first have to travel through the mantle. Before being detected at a seismic observatory, it will then have to pass through the fluid OC (and perhaps the solid IC) and then through the mantle again. In order to study conditions at the CMB, Creager and Jordon (1986) used travel times of PKIKP (waves passing nearly vertically through the CMB, OC and IC) and PKP (waves spending longer in the mantle and passing through the OC only) (Figure 2.15). They

. suggested three possibilities for the source of the observed anomalies – the D'' region above the CMB, bumps on the CMB, and a thin, highly heterogeneous layer in the fluid OC beneath the CMB. They reject the D'' layer – as did Morelli and Dziewonski (1987) in their analysis described above. They reject the second possibility because of the large range of undulations of the CMB (± 8 km), although for their PKP data alone the range of topography of the CMB is similar to that found by Morelli and Dziewonski. They therefore prefer their last possibility – large scale heterogeneity in a thin boundary layer at the top of the OC.

Although boundary layer models involving large scale variations in OC chemistry have not been fully investigated, it seems difficult to sustain significant lateral density variations in the OC. The fluid OC can flow as easily as water (it has about the same viscosity) and it is doubtful whether the required heterogeneity could be maintained. It seems far more likely that the anomalies in the seismic data are due to undulations of the CMB, including variations in the composition of the mantle near the CMB – possibly lumps of rock formed elsewhere that have settled there. The core mantle boundary will be considered again in Chapter 6.

Lay and Young (1990) have examined the structure of the outermost 200 km of the OC using differential travel times SKS–SKKS. They found that the best fit to their data is a model in which the P wave velocity in the outermost 50–100 km of the OC is 1–2% slower than in standard Earth models. Such an anomalous region could have arisen from chemical stratification during core formation, from segregation of a light component in the core, or from chemical reactions between the mantle and core. Lay and Young stress that their interpretation is not unique, but the existence of such an OC boundary layer would have profound implications for models of the geodynamo (section 5.3), and chemical and dynamical interactions between the core and mantle. There can be no vertical flow in the layer if it is to persist, although there could be strong horizontal flow velocities induced by variable heat flux on the CMB as a result of temperature variations in the D'' layer.

2.7 THE INNER CORE (IC)

Morelli *et al.* (1986) used PKIKP travel times to investigate the structure of the IC. They found that the average velocity of P waves along the Earth's rotational axis is about 1% faster than in the equatorial plane and concluded that the IC behaves as an anisotropic medium with cylindrical symmetry aligned along the rotational axis. Similar conclusions were reached by Giardini *et al.*,

(1987) from an analysis of the anomalous splitting in free oscillation spectra. On the assumption that the IC is isotropic, variations of several per cent in IC properties and undulations of 8 and 23 km on the CMB and ICB respectively are required, which seems very unlikely. They therefore preferred the alternative hypothesis that the IC is anistropic.

Shearer *et al.* (1988) also noted that seismic waves passing through the middle of the Earth travel slightly faster in a north to south direction than in an east to west direction and concluded the best explanation is IC anisotropy. However they prefer a uniformly anistropic IC model, unlike that of Morelli *et al.* (1986) which concentrated the anomaly near the surface of the IC. Their model does admit however the presence of some lateral heterogeneity in the IC. Shearer and Toy (1991) in a later study concluded that both anisotropy and heterogeneity may well be present since any pattern of convection within the IC, which gives rise to anistropy through preferred crystal alignments, must also result in some heterogeneity.

The mechanism by which the IC grew with such a preferential alignment is not at all clear. Morelli *et al.* pointed out that the IC consists of Fe in its hexagonal closely packed ε phase, a crystallographic form with cylindrical anisotropy. Moreover crystallization of the IC would have taken place in the presence of fluid motion, electric currents and electro-magnetic fields, all of which are coupled to the Earth's rotation and it is possible that this could result in orientation of the crystals along the Earth's rotation axis. Sayers (1990) has shown that an anisotropic IC with cylindrical symmetry aligned along the Earth's rotational axis places an important constraint on its crystal structure. In particular he has shown that it is inconsistent with a cubic crystal structure, providing further evidence that the IC consists of Fe in its hexagonal closely packed ε phase.

Jeanloz and Wenk (1988) have put forward an alternative explanation *viz* that the IC is convecting and that the anisotropy is induced by the flow. They estimated the Rayleigh number* of the IC to be 10^{10}. Even allowing for an uncertainly of four orders of magnitude, this is at least 100 times greater than the critical value. The high value of Poisson's ratio and the relatively low Q of the IC suggest that some of the region may be partially molten. If this were the case, it would strengthen the suggestion that the IC is sufficiently inviscid to permit convection.

---

*The Rayleigh number is the ratio of buoyancy to diffusive effects. Convection sets in at a critical value of the Rayleigh number. The larger the Rayleigh number, the more vigorous the convection.

Souriau, A. and Souriau, M. (1989) have used differential travel times and relative amplitudes of PKiKP (waves reflected from the ICB) and PcP (waves reflected from the CMB) at nearly vertical incidence to estimate the ellipticity and density at the ICB. This approach enables them to minimize contributions from the source and mantle, and, provided the liquid OC has no significant lateral heterogeneity, yields information depending only on the configuration of the CMB and ICB. They found that the ICB is close to sphericity with at most a small flattening.

Velocity and density contrasts at the ICB are important parameters which constrain the chemical and physical state of the IC. As will be discussed in Chapter 3, the composition of the IC is usually assumed to be pure Fe, although there has been some suggestion that, like the OC, it might contain some lighter elements (e.g. Jephcoat and Olson, 1987). This is important since, if true, it would change the amount of energy released during the growth of the IC which is available to drive the geodynamo (section 3.5). If the IC were pure solid Fe in the hexagonal close packed ε phase, shock wave data indicate the density jump $\Delta \rho$ at the ICB to be $1.63 \, \mathrm{g \, cm}^{-3}$ and static high pressure data a $\Delta \rho$ of $2.07 \, \mathrm{g \, cm}^{-3}$ (Jephcoat and Olson, 1987). Souriau and Souriau (1989) examined the amplitude ratios PKiKP/PcP and estimated $\Delta \rho$ to be $1.35-1.66 \, \mathrm{g \, cm}^{-3}$. These estimates are considerably above the maximum value of $0.5-0.6 \, \mathrm{g \, cm}^{-3}$ obtained from free oscillation data by Widmer *et al.*, (1988) and above the PREM value $(0.6 \, \mathrm{g \, cm}^{-3})$.

Shearer and Masters (1990) have also looked at PKiKP amplitudes and concluded from PKiKP/PcP amplitude ratios that $\Delta \rho$ is less than $1.0 \, \mathrm{g \, cm}^{-3}$ and that the shear velocity at the top of the IC is greater than $2.5 \, \mathrm{km \, s}^{-1}$. They also analysed about 50 free oscillation data that were particularly sensitive to the structure of the IC. They found that $\Delta \rho$ at the ICB cannot be significantly different from $0.55 \, \mathrm{g \, cm}^{-3}$ and that the average shear velocity in the IC is $3.45 \pm 0.1 \, \mathrm{km \, s}^{-1}$. The free oscillation data are thus compatible with their PKiKP amplitude bounds. If these lower values of $\Delta \rho$ are substantiated, it would further strengthen the suggestion that the IC is not pure Fe and, like the OC, does contain some lighter elements. This question is discussed further in section 3.5.

3

The constitution of the Earth's core

3.1 INTRODUCTION

Before ultra-high pressure experiments using shock wave techniques and the diamond-anvil cell (to be described later), one could but hypothesize on the behaviour of materials below the crustal layers. The dangers of such hypotheses and perhaps self-delusions based on an over-simplification of conditions that exist at depth within the Earth have been stressed by Professor Birch (1952).

Unwary readers should take warning that ordinary language undergoes modification to a high pressure form when applied to the interior of the Earth: a few examples of equivalents follow:

High pressure form	*Ordinary meaning*
certain	dubious
undoubtedly	perhaps
positive proof	vague suggestion
unanswerable argument	trivial objection
pure iron	uncertain mixture of all the elements.

Most information on the properties of the materials below the crustal layers has been obtained in the past by theoretical interpretation of phenomena observed at the Earth's surface, while at great depths extrapolation of theoretical equations is frequently needed far beyond the conditions under which they were originally developed. An appreciation of many geophysical problems often requires a synthesis of knowledge from many disciplines such as astronomy, chemistry, geology, mathematics and physics. It is impossible for anyone to keep abreast of recent developments in any one of these fields, let alone in all of them. Thus a geophysicist may not be familiar with the underlying assumptions and limitations of a certain result he is using with the result that his extrapolation or extension of the theory may be grossly in error. To quote Professor Gutenberg (1959),

> conclusions concerning the deep portion of the mantle and the core may be subject to two major sources of error: those resulting from misinterpretation of observations, and those from application of theoretical equations which fit the problem only poorly or not at all and, in addition, may contain incorrectly estimated numerical factors.

Moreover, scientists often have to make use of results in a field with which they are not familiar. Frequently they quote as well-established facts the results for 'models' which had only been proposed as a working hypothesis. To quote Professor Gutenberg again,

> It is nothing unusual that a tentative suggestion or model of a geophysicist is quoted by a writer in a different field as a proof for one of his hypotheses – creating the impression that such a tentative suggestion is in fact a definite result.

The mathematical development of a theory is frequently carried out to a much greater accuracy than is warranted by the imperfectly known or questionable assumptions of the physics of the process. Proper statistical analysis is essential in an interpretation of the results of many geophysical measurements. However, if 'probable errors' are calculated (by a least squares analysis or otherwise), then it does not necessarily follow that, if small, the result indicates confirmation of a theory – it may merely indicate good agreement between observations or the reproducibility of some experimental technique. Systematic misinterpretation of the results or incorrect assumptions may produce actual errors far greater than any calculated 'probable errors'.

With these words of warning, let us now consider the constitution of the Earth's core. A fundamental problem is to determine an equation of state (EOS) by which we mean a relationship between the pressure, density and temperature of a material. An equation of state cannot involve the history of the material and thus non-hydrostatic stresses are excluded – large non-hydrostatic stresses in general lead to irreversible (plastic) deformation. The pressures and temperatures in the deep interior of the Earth are on the one hand sufficiently high to make them difficult to reproduce in the laboratory, and on the other hand sufficiently low to make theoretical quantum statistical models not applicable.

3.2 EXPERIMENTAL METHODS

Experimental results of wave propagation show that the velocity of compressional waves depends mainly on density and mean atomic weight. Most common rocks have mean atomic weights around 21 or 22 irrespective of composition, so that rocks and minerals of very different composition may have the same densities and seismic velocities. Thus seismic data alone are not of much help in inferring chemical composition and laboratory experiments at high temperatures and pressures comparable to conditions deep within the Earth are necessary as well. The early experimental work was carried out by Bridgman. However the highest pressures he was able to attain were about 100 kbar, equivalent to a depth of only 300 km in the Earth*

The pressure and temperature at the CMB are 130 GPa and about 4000 K respectively. The combination of high temperatures and pressures can be generated under dynamic compression (shock wave) techniques as will be discussed below. The creation of these conditions in static experiments however has potentially more flexibility in terms of control of temperature and pressure, which can be changed independently. Moreover the experiments can be carried out for almost any length of time. Static experiments will be considered first.

3.2.1 Static Measurements

The great advances over Bridgman's static experiments are due to Mao and Bell who developed a diamond-anvil high-pressure cell at the Carnegie Institution, Washington. A description of the apparatus with continuing improvements in design and results have been published regularly in the Annual Reports of the Director of the Geophysical Laboratory from the year 1974–1975 onwards. Pressures of 1.2 Mbar have been reached and held constant for two weeks enabling studies of the region near the CMB to be made. Higher pressures have now been attained, well in excess of that at the centre of the Earth.

Higher pressures may be achieved either by increasing the force or decreasing the area. The conventional way of trying to produce the high pressures comparable to those in the deep Earth was to build bigger and bigger presses. With cubic centimetre sized

---

* The SI unit of pressure is the pascal (Pa), but much experimental work at high pressures still quotes results in kilobars (kbar). The conversion is 1 giga pascal (GPa) = 10 kbar.
Pressures in the Earth's core exceed 100 GPa = 1 Mbar.

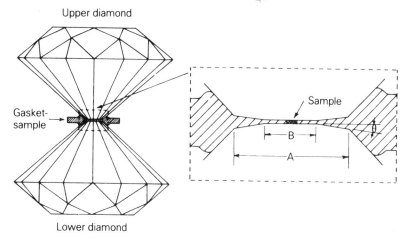

Upper diamond

Gasket-sample

Lower diamond

Sample

Figure 3.1 Sketch of the diamond-anvil pressure cell. Distance from the top surface of upper diamond to bottom surface of lower diamond ~5 mm. Inset – magnified view of a cross-section of the gasket-sample assembly. The sample width is 250 μm (after Mao and Bell, 1978).

samples, the largest presses could not create pressures ≳0.3 Mbar. Instead of increasing the force applied to the sample to try and reach higher pressures, Mao and Bell reduced the area over which the force was applied. This is essentially the idea behind the diamond-anvil cell.

The basic principle is very simple. A sample is placed between the flat, parallel faces (about 350 μm across) of two opposed diamond anvils and is subjected to pressure when a force pushes the two anvils together. (Figure 3.1). The force can be applied by hand and multiplied by a nut-cracker like device. When cranked up to Mbar pressures, the cell can be hand-carried across the laboratory or even across the country. A variety of sophisticated measurements can be carried out on materials of microscopic dimensions. The high pressures needed for studying the structure and possible phase transformations of minerals are thus possible, since very small samples (dimensions <10 μm) can be examined. Moreover single crystals can be preserved with gas pressure-transmitting media to pressures >50 GPa.

Below about 1200 K, the anvils can be heated resistively. For higher temperatures, the samples in the cell can be heated with an intense laser source. The diamond anvils are transparent to visible radiation and, depending on the type of diamonds used, transparent to much of the ultraviolet and infra-red spectrum as well. Thus only the sample is heated, avoiding any excessive heating of the diamonds and supports. Many minerals absorb near infra-red radiation and can be heated to temperatures far in excess of those in the core with a sufficiently powerful laser. The coupling of the near infra-red radiation is particularly strong for iron bearing phases, important for simulation of the mantle and the core. To heat samples that may be transparent at this wavelength, strongly absorbing material such as graphite may be mixed in with the sample. The temperature of the sample is measured with an optical pyrometer.

The ultra-high pressures generated within a diamond-anvil cell must be determined by *in situ* methods. This is more difficult than temperature measurements, since the pressure is not hydrostatic (the material tends to be extruded from between the anvils). A widely-used technique is the ruby fluorescence method which involves measuring the pressure shift of the R_1 luminescence band of ruby which shifts to longer wave lengths as pressure increases. A tiny chip of ruby 5–10 µm in dimension is placed in the pressure medium along with the sample and the fluorescence is excited with a He–Cd laser line or any source of strong light. The R lines of the ruby are quite intense and the doublet R_1 and R_2 has wavelengths of 6927 and 6942 Å respectively at atmospheric pressure. Under pressure these shift to higher wavelengths, the shift being approximately linear with pressure. There is a small positive non-linearity in the pressure dependence of the wavelength shift of the R lines – the linear scale underestimates the pressure (<3% at 1 Mbar).

However it is necessary to calibrate the R_1 scale directly in the higher range of pressures and this can only be obtained from independently determined pressure-volume relationships of suitable materials. Mao and Bell (1976) used 4 metals – Cu, Mo, Pd and Ag – for which accurate pressure-volume relationships from shock measurements up to 1 Mbar were available. The calibration has been extended to 180 GPa with samples consisting of ruby and either gold or copper (Bell *et al.*, 1986).

In other experiments, Bell and Mao (1979) made measurements of the force per unit area to compare with the accurate ruby-fluorescence pressure-calibration scale to 1 Mbar and to provide calibration above 1 Mbar. In these experiments, ruby crystals located throughout the surface of a stainless-steel sample were used with the fluorescence technique to determine the pressure distribution and the mean pressure. The force-per-unit-area measurements are less precise than the ruby measurements, but are considered absolute pressure measurements. The two independent calibrations were done simultaneously and yielded the same results, confirming the ruby pressure scale to 1 Mbar.

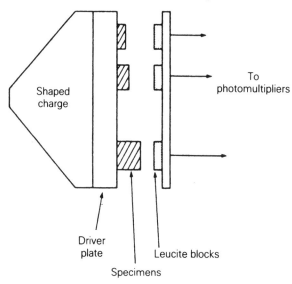

To
photomultipliers

Shaped
charge

Driver
plate

Leucite blocks

Specimens

Figure 3.2 Generation of shock waves in solid specimens (after Cook, 1980).

Mao *et al.* (1985) later redesigned the diamond cell and obtained pressures up to 2.8 Mbar at the centre of the diamond face, equivalent to that well inside the Earth's IC. Ruby-fluorescence is not observed above about 1.85 Mbar and pressures had to be determined by extrapolation from lower values. The pressure calibration was done indirectly from load calculations. The shift of the fluorescent line of ruby crystals placed in the sample could not be used because of strong interference from diamond–anvil fluorescence at pressures above 2.7 Mbar. Xu, Mao and Bell (1986) have since reached pressures of 5.5 Mbar, well in excess of that at the centre of the Earth (3.5 Mbar). In these experiments, the overlapping diamond emission was found to disappear at pressures above 3 Mbar, and the ruby pressure calibration scale could be employed once again.

3.2.2 Dynamic Measurements

If an explosive attached to one face of a slab of material is detonated, a high pressure wave passes through the slab. Better results may be obtained if the explosive charge drives a metal plate against the face of the sample rather than attaching the explosive directly to the sample. The charge is so constructed that the impulse from the explosion is applied simultaneously over the whole face of the sample. (Figure 3.2). If the pressure applied to the slab by the impact is sufficiently great, the material of the slab will no longer deform elastically and a shock wave will be propagated through the slab at a velocity greater than the speed of elastic waves. The pressure rises in a thin layer of material to the value set up by the explosive and, behind the thin layer, the material as a whole is set in motion. When the shock reaches the far side of the slab, it is reflected as a rarefaction wave, satisfying the condition that the pressure on that side is atmospheric (i.e. $\simeq 0$). The whole slab is then moving at the speed of the material behind the shock. The equations of the conservation of mass, momentum and energy together with the measured velocities of the shock wave and the bulk material enable the pressure, density and internal energy of the shocked material to be calculated and hence an equation of state (EOS) to be set up.

If (p_0, v_0) and (p_1, v_1) are the pressure and specific volume (i.e., the reciprocal of the density) before the shock arrives and after the shock has passed, then it can be shown that

$$\frac{v_1}{v_0} = \frac{u_s - u_p}{u_s} \tag{3.1}$$

and

$$p_1 - p_0 = \frac{u_s u_p}{v_0} \tag{3.2}$$

where u_s and u_p are respectively the velocity of the shock wave and that of the slab after the shock has passed. The essential measurements thus consist of the determination of the velocity u_s of the shock and the velocity u_p of the material behind the shock.

Both optical and electronic techniques are used to measure shock propagation times through samples and hence shock velocity. The electronic method uses electrical contactor pins. These are basically switches which are closed by the particle velocity induced by the shock wave, or the shock pressure which induces electrical conductivity in a thin insulating layer. There are two optical methods – the flash-gap method and the reflected light method. The flash-gap method employs a series of leucite blocks set about 0.01 mm above the surface of which the shock wave arrival is to be detected. This 0.01 mm thick space is filled with argon or xenon. Blocks are placed on either side of a sample to record shock entrance into the sample and on the upper surface of the sample to detect the completion of the shock through the sample. On shock arrival and impact of the sample or driver plate against the gap, the gas trapped

between the leucite blocks is shock heated to temperatures of
~30 000–50 000 K. The light from the shocked gas is recorded
photographically on a streak camera. An advantage of this method
is that data for many samples may be obtained in a single experiment. The other optical method uses reflected light recorded with
a streak camera. In this case the image of a bright-light source
reflected by the polished driver plate, sample surface, or mirror in
contact with these is recorded as a function of time. The change of
light intensity due to a roughening of the free surface when the
shock wave arrives or a tilting of the free surface may be used as a
shock wave detector. A detailed description of the experimental
techniques has been given by Ahrens (1987).

The composition of the explosive charge determines the speed of
the plate and hence the strength of the shock. Each shock of given
strength thus gives one point specified by p and v in the shock. If
material is given a series of shocks of different strengths, then a
curve of density against pressure called the Hugoniot can be built
up. It is neither an isotherm nor an adiabat, for the internal energy
and the temperature vary from point to point along the curve. On
an isotherm or an adiabat, successive states can be obtained by
following the curve. Successive states during the Hugoniot cannot
be achieved one from another by a shock process. The Hugoniot
curve just represents the locus of final shocked states. A major
problem in shock wave experiments is the deduction of an isothermal EOS from the Hugoniot.

The Hugoniot curves, upon reduction, yield pressure-
temperature states for different materials. One form of the EOS
consists of pressure-density isentropes (constant entropy curves).
Upon differentiation the isentropes yield the seismic parameter

$$\phi = \left(\frac{\delta p}{\delta \rho}\right)_s$$

Direct comparison is thus possible between values of ϕ obtained
from the seismic velocities V_p and V_s (equations 2.9 and 2.10) and
values of ϕ measured for rocks and minerals in the laboratory under
similar conditions of temperature and pressure.

Two assumptions are made in shock wave experiments – that the
measured states are in thermodynamic equilibrium and that the
compression for a given pressure is the same as that which would
be produced by a hydrostatic pressure of the same magnitude. The
first condition is satisfied if thermodynamic equilibrium is attained
in about 10^{-7} s or less. This implies a shock front with a thickness
of a few tenths of a millimetre or less.

Figure 3.3 is a plot of $\phi^{1/2}$ versus density along the Hugoniot

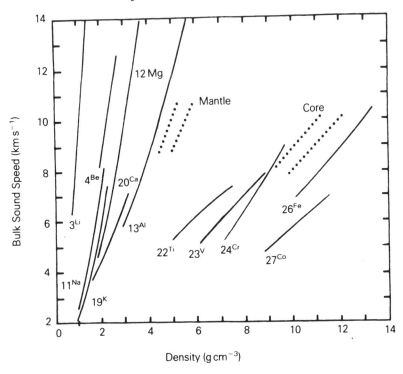

Figure 3.3 Bulk sound speed

$$\left(\frac{\partial p}{\partial \rho}\right)_s^{1/2}$$

versus density for a number of elements along the Hugoniot compression curves. Atomic numbers are attached to each curve. The areas in which the corresponding quantities for the Earth's mantle and core must lie are indicated by the pairs of dotted lines (after Birch, 1968).

compression curves for a number of elements. Atomic numbers are attached to each curve. The areas in which the corresponding quantities for the Earth's mantle and core must lie are indicated by the pairs of dotted lines. Figure 3.4 shows density models of the core compared with shock wave data for iron.

These figures indicate that both densities and bulk moduli in the OC are less than those of Fe under equivalent conditions, although their gradients through the OC are consistent with gross chemical homogeneity (i.e., uniform intermixing of Fe with a lighter, more compressible element or compound). Both densities and bulk moduli for the IC are compatible with those of Fe, suggesting that

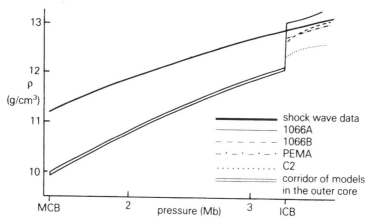

Figure 3.4 Density models of the core compared with shock wave data for iron (after Jacobs, 1980).

the inner-core boundary (ICB) is likely to be a compositional as well as a phase boundary, in which case the boundary cannot be used as a fixed temperature point on the melting curve of Fe (or a related compound).

The reason for the OC being molten is simply that it is alloyed with some lighter component such that not only its density but also its melting point is lowered below that of pure Fe. Jeanloz (1979) showed how data on bulk modulus and density provide constraints on the composition of the OC. Figure 3.5 gives the unique combination of density and bulk modulus as a function of mass fraction of the core that the lighter compound X must possess to satisfy the seismic data for the OC as well as the shock wave data on pure Fe.

3.3 THE ABUNDANCE OF THE ELEMENTS

The abundance of the elements in the universe has been obtained from the spectographic analysis of light from the sun and stars. Light from the sun may be split into different colours or wavelengths using a prism or diffraction grating. The spectrum shows a number of dark lines running across it – the same lines can be produced in the laboratory when different elements are put in a flame. The reason for this is that different elements absorb or emit light corresponding to a characteristic and precise wavelength. The wavelengths of the lines in the spectrum from the sun tell us what elements are present in its outer layers where atoms absorb some of the light coming from the interior. The strengths of these lines then

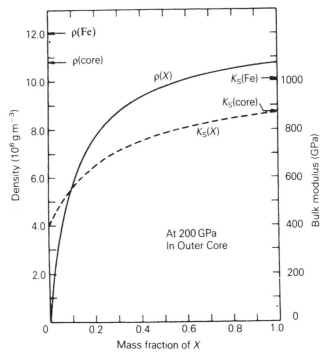

Figure 3.5 Diagram showing the values of density ρ and bulk modulus K_s that must simultaneously be satisfied by a candidate compound X coexisting in a given mass fraction with Fe in the OC at a pressure of 200 GPa. Errors in density may be as much as ±3 − 5%, those in bulk modulus may be up to ±5 − 10%. The densities and bulk moduli of the core and of iron (under the same conditions) used to construct the figure are also shown (after Jeanloz, 1979).

indicate the amount of each element that is present. The two lightest elements, H and He make up more than 99% of the visible universe, the other elements showing a rapid exponential decrease (Figure 3.6). On the Earth H and He are much rarer, since being gases, they largely escaped into space when the Earth was formed.

Further information on the abundance of the elements has come from the analysis of meteorites – particularly Type I carbonaceous chondrites (CI chondrites) which are thought to contain material left over from the formation of the solar system. The chemical composition of these meteorites agrees very closely with that of the sun, failing only for the very volatile elements such as H, C, N, O, the rare gases and the light elements Li, Be and B (Figure 3.7). As a result CI chrondrites are usually considered as representative of the average composition of the solar system, including the Earth.

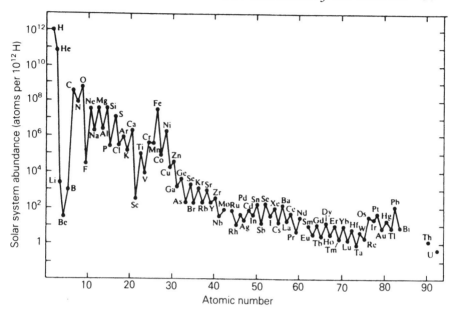

Figure 3.6 Solar system abundances of the elements, showing the relative number of atoms present on a logarithmic scale, normalized to the value 10^{12} for H (based on a combination of solar spectrum and meteorite data) (after Cox, 1989).

The composition of the corona and photosphere of the sun have recently been redetermined by Breneman and Stone (1985). Their results are based on the solar energetic particles observed during large solar flares by the spacecraft Voyager 1 and 2 and are in general agreement with recent spectographic values. The Fe abundance is about 40% higher than earlier measurements and Fe and Ca relative to Al, Si and Mg are 30–40% higher than chondritic values. Anderson (1989) thus suggests that chondritic meteorites may not be representative of the solar nebula, and that the terrestrial planets may be richer in Fe and Ca than previously supposed.

The abundance of the elements in the Earth tells a different story. For the Earth as a whole, the most abundant elements are Fe, O, Si, Mg, Ni, S, Ca and Al accounting for more than 99% of the mass of the Earth. In the crust the most abundant elements are O, Si, Al, Fe, Mg, Ca, K and Na. Most of the Fe has sunk to the core, the crust consisting almost entirely of O compounds, especially silicates of Al, Ca, Mg, N and K. The average composition is in effect that of igneous rocks.

It should be noted that differentiation of the elements in the Earth did not lead to a vertical arrangement based entirely on their

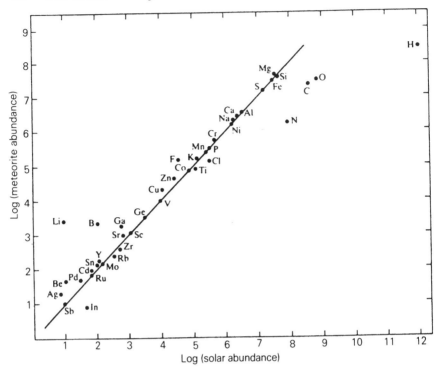

Figure 3.7 Comparison between abundances of elements with atomic numbers between 1 and 51 found in carbonaceous chondrite meteorites, with those in the sun (after Cox, 1989).

relative weights. Properties such as melting points, densities and chemical affinities of the compounds they formed governed the distribution of the elements, rather than the properties of the elements themselves. Some heavy elements, like U and Th have strong tendencies to form oxides and silicates which are light and could rise up into the crust.

Some of the elements that have long been known to and used by man are actually quite rare, e.g., Cu is less abundant than Zr and Pb is comparable in abundance with Ga. Conversely there is the relative abundance of many unfamiliar elements. Thus the abundance of Rb is comparable to that of Ni, and V is much more abundant than Sn. The answer to this seeming paradox is the distinction between abundance and availability. Some elements, although present in the crust in considerable amounts, are systematically dispersed throughout common minerals and never occur in any concentration.

3.4 THE CONSTITUTION OF THE OUTER CORE (OC)

There is no firm evidence as to what is the light element in the OC. It must be reasonably abundant, miscible with liquid Fe and possess chemical properties that would allow it to enter the core. Possible candidates are H, He, C, O, N, Mg, Si and S. The composition of the core has been reviewed by Brett (1976, 1984), the most favoured candidates for the light element being S or O. Si was long the preferred choice, but has now been rejected. Ringwood (1966) championed Si for a number of years, but in the face of a number of difficulties arising from models of the accretion of the Earth and subsequent core formation that Si entailed, Ringwood abandoned it and now favours O. Shock wave measurements on SiO_2 are also not favourable for Si being the major light element in the core and most people now favour S or O. Jeanloz (1990) lists all possible candidates for the light alloying element and gives the arguments both for and against them.

3.4.1 Sulphur in the outer core

Originally, sulphur was proposed as the light component in the OC by Murthy and Hall (1970) primarily because it is depleted in the rest of the Earth compared to cosmic abundances. If sufficient S were present in the core, the Earth as a whole can be undepleted in S. Ahrens *et al.* (1979) have carried out shock wave experiments on pyrrhotite ($Fe_{0.9}S$) over the pressure range 0.03 to 1.58 Mbar. If the data are reduced to isotherms, a nearly constant S content of 9–12% throughout the OC satisfies the seismic density-pressure curve. Later Brown *et al.* (1984) extended the data up to a pressure of 2.74 Mbar. If S is the light element in the core, their results are consistent with a homogeneous mix containing 10 ± 4 wt% S. Kraft *et al.* (1982) obtained similar results using static high pressure data. The experimental data thus indicate that not nearly enough S is in the core to make up for the deficiency of the bulk Earth. Although the original argument of Murthy and Hall is not valid, it does not rule out S as the lighter component in the core. FeS is a good electrical conductor and melts at temperatures several hundred degrees below the melting point of mantle minerals, making it an ideal component for the OC.

Williams and Jeanloz (1990) have investigated the melting temperatures of FeS (troilite) and of a 10 wt% sulphur-iron alloy to pressures of 120 and 90 GPa respectively, using a laser heated diamond anvil cell. They found that FeS melts at a temperature of 4100 ($\pm$300) K at the pressure at the CMB. The melting

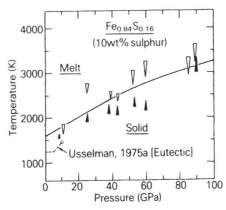

Figure 3.8 Bounds on the melting temperature of a 90/10 wt, % iron-sulphur mix. Solid triangles represent the highest temperature measured within solid $Fe_{0.84} S_{0.16}$ at the indicated pressure, open inverted triangles represent the lowest temperature measured within liquid $Fe_{0.84} S_{0.16}$. Lengths of triangles represent uncertainties in temperatures (after Williams and Jeanloz, 1990).

temperature for the 10 wt% sulphur model of the core is lower than that of both Fe and FeS up to pressures of at least 90 GPa (Figure 3.8). This eutectic-like behaviour is very different from that in the Fe–FeO system which Knittle and Jeanloz (1991a) found shows solid-solution like properties above about 70 GPa for compositions with 10 wt% oxygen. If the only alloying elements of iron in the OC are sulphur and oxygen and the OC is entirely liquid, Williams and Jeanloz (1990) estimate the minimum temperature at the top of the core to be 4900 (±400) K.

One of the main differences between the O and S hypotheses is that under the oxide hypothesis, the core must have acquired its present composition at high pressures – below the metallization pressure, O does not combine with Fe in any significant amount. In contrast S can readily be alloyed with Fe at low pressures. If the formation of the core took place early in the Earth's history so that differentiation took place at the same time as most of the accretion, then the core started forming at a relatively low pressure which could lead to difficulties with the hypothesis that it is an alloy of Fe and O.

The result of combining either S or O with Fe is to lower the melting point of the compound. S has a much larger effect than O in this regard so that melting would begin at significantly lower temperatures in a sulphide composition than it would in an oxide one. Thus again it might seem easier for a core to start forming if it is S rich rather than if it were O rich.

3.4.2 Oxygen in the outer core

The idea that oxygen might be the principal light alloying element in the Earth's OC was revived by Ringwood (1978) who argued that the solubility of FeO in liquid Fe might increase rapidly with both temperature and pressure. Jeanloz and Ahrens (1980) carried out shock wave measurements on wüstite ($Fe_{0.94}O$) to pressures over 1.5 Mbar. Their data show that the density of the OC is equal to that of an equal mix (by weight) of Fe and FeO (~10 wt% O).

The extent to which FeO is soluble in molten Fe is determined primarily by the nature of the bonding in the competing crystalline FeO phase. There has been much controversy over the interpretation of the Jeanloz-Ahrens data which showed a phase transition at about 0.7 Mbar. The situation is further confused since X-ray diffraction studies on wüstite up to 120 GPa at room temperature using a diamond-anvil cell found no discontinuous volume reduction in contrast to the large density increase at 70 GPa observed in the shock wave experiments.

McCammon *et al.* (1983) have re-examined the process whereby FeO may be segregated from a silicate oxide into a core liquid. They showed that extrapolation of the solubility data of O in liquid Fe indicates that the liquid miscibility gap in the system Fe–FeO at atmospheric pressure should close at about 4250 K. At high pressure (about 90 GPa), the phase diagram of Fe–FeO is qualitatively similar to the system Fe–FeS displaying complete miscibility and a eutectic temperature well below the melting point of pure Fe.

Ohtani and Ringwood (1984a) have carried out experiments on the solubility of crystalline FeO in molten Fe at temperatures between 2100 and 2550 C, and concluded that complete miscibility should be attained at temperatures around 2800 C. Ohtani *et al.* (1984b) also carried out experimental and theoretical investigations of the effect of pressure on the solubility of FeO in molten Fe. Solubility increases sharply with pressure, the liquid immiscibility gap contracts and disappears around 20 GPa. They predicted that the Fe–FeO phase diagram should resemble a simple eutectic system above about 20 GPa.

These results have been reinforced by Knittle and Jeanloz (1986) who determined the phase diagram of FeO up to pressures of 155 GPa and temperatures of 4000 K using shock wave and diamond-anvil cell data. At ambient conditions, FeO is nonmetallic and forms an immiscible liquid with metallic Fe. However metallization of FeO would allow O to alloy with Fe and thus provide a mechanism for incorporating O into the core. Knittle and Jeanloz found a metallic phase of FeO at pressures greater than

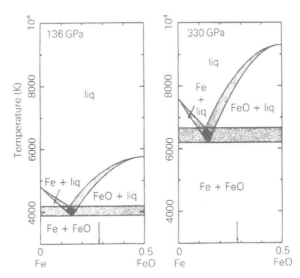

Figure 3.9 Suggested phase diagrams for the Fe-rich portion of the ε–Fe–O system at 136 and 330 GPa. The stippled regions show the variability in the position of the phase boundaries, and the cross-hatched region shows the variability in the position of the eutectic. The arrow at the bottom of each figure represents the bulk composition of the core (assuming the IC is pure Fe) calculated from the estimate of Jeanloz and Ahrens (1980) for the amount of O in the OC, based on the assumption that O is the only light element in the OC (after Anderson *et al.*, 1989).

70 GPa and temperatures above 1000 K. The high pressure necessary for this metallization suggests that the composition of the core may not have been established until a relatively late stage in the evolution of the Earth. Knittle and Jeanloz estimated that sulphur may have been an initial component of the primordial core and that the Earth grew to about 50–70% of its final radius (10–35% of its final mass) before O began to alloy with the metal.

Contradictory results have been obtained by Anderson *et al.*, (1989) who constructed phase diagrams for Fe–O–S systems at core conditions based on experimental data and theoretical arguments. They found that a solid solution exists between ε Fe and S at high pressures, which is not the case for the ε Fe–O system. Experimental data (at pressures greater than 100 GPa) show that FeO melts at higher temperatures than either Fe or O. This suggests that solid FeO remains a stable compound, with a possible miscibility gap between FeO and Fe. Theoretical interpretation of the melting curve of FeO indicates that this behaviour should continue throughout the pressure range in the Earth's core. Hence the Fe-rich portion of the Fe–O phase diagram is predicted to have a eutectic between εFe and Fe–O at core pressures. Figure 3.9 shows that an Fe–O liquid composition of OC density falls well to the O-rich side of the Fe–FeO eutectic. Hence the solid in equilibrium with the liquid component of the core would be Fe–O rather than Fe. At pressures in the IC, FeO has a density of 10.5–11.0 g cm^{-3} well below densities based on seismic data of 12.7 to 13.0 g cm^{-3}. This would lead to a core where the solid phase is less dense than the liquid phase. Thus O cannot be the only light element in the core. This difficulty does not exist with S, since the Fe–S solidus is always more Fe-rich than the co-existing liquids.

It must be stressed that all these models of the core are based upon binary phase diagrams. The core may well contain significant amounts of both O and S, together with smaller amounts of other elements. Agee (1990) has carried out high pressure (up to 26.5 GPa) melting experiments on the Allende CV3 carbonaceous chondrite and found that FeO-rich magnesiowüstite (Mg, Fe)O is an abundant crystallizing phase at temperatures near the Allende silicate liquidus. If the Earth accreted from unfractionated primitive meteoritic material similar to that of carbonaceous chondrites and later passed through a high temperature molten stage, then during cooling and crystallization, FeO-rich magnesiowüstite could be segregated into the Earth's deep interior. Agee thus suggests that magnesiowüstite fractionation could have depleted the initial FeO content of the Earth's chronditic mantle and added to the Fe–Ni–S proto-core.

3.4.3 Potassium in the outer core

Another question that has provoked much discussion is whether there are appreciable amounts of K in the core. It was suggested (e.g., Lewis, 1971; Hall and Murthy, 1971) that in the deep interior of the Earth under the strongly reducing conditions prevailing at the time of Fe–FeS liquid segregation, the alkali elements K, Rb and Cs would show strong chalcophilic* tendencies in possible reactions between silicate and Fe–FeS liquids. The difference between the terrestrial and chondritic ratio K/U has also been ascribed to the partitioning of K into the core. The presence of ^{40}K in the core would have profound implications for the thermal history of the Earth. The heat generated by the decay of ^{40}K in the liquid OC would also set up convective motions necessary to power the geodynamo (section 5.3).

---

*Chalcophilic – having a strong affinity for sulphur.

Oversby and Ringwood (1972) investigated the distribution of K in meteorites and carried out experimental work on the distribution of K between a synthetic basalt and an Fe, FeS metallic phase containing 28% S. They concluded that at most 2% of the total amount of K in the Earth would be in the core. Goettel and Lewis (1973) disputed their results arguing that the chemical and physical conditions that are relevant to the partitioning of K between an Fe–FeS core and the silicate mantle and crust are those that existed during the primary differentiation of the Earth into core and mantle and not the conditions in the present crust and mantle. However further partitioning experiments by Ganguly and Kennedy (1977) and experiments on the melting relations of the Allende chondrite by Seitz and Kushiro (1974) confirm Oversby and Ringwood's conclusions. The question seems now to have been settled by Somerville and Ahrens (1980) who carried out shock-wave experiments on $KFeS_2$ up to pressures of 110 GPa and found no evidence for any appreciable amounts of K in the core.

Murrell and Burnett (1986) have carried out experiments on the partitioning of K, U and Th between sulphide and silicate liquids at 1450 C and 1.5 GPa and also at 1 atmosphere. They concluded that U and Th partitioning into Fe–FeS liquids would be far more important than K partitioning. Gubbins (1981) estimated the power needed to drive the geodynamo to be about 10^{13} W. This would require a concentration of about 150 ppm of K in the core. Murrell and Burnett estimated that only 50 ppb of U or 180 ppb of Th would be required. They also concluded that pressure effects near the CMB would need to increase the partition coefficent of K by a factor of about 10^3 with a much smaller increase in that of U in order for terrestrial and chondritic abundances of K and U to be the same. It must be pointed out that the interpretation of their data is only valid if S is the most important light component in the core.

3.5 THE CONSTITUTION OF THE INNER CORE (IC)

By comparing densities of the IC determined from seismology with those obtained from static and shock wave experiments on Fe and FeS at high pressures, Jephcoat and Olson (1987) concluded that the IC is not pure Fe, i.e., like the OC it must contain some light component as well. They estimated that 3–7 wt% S is required. The incorporation of light components into the IC would have a direct bearing on the driving mechanism of the dynamo that produces the geomagnetic field (section 5.3). The most favoured method for driving the geodynamo is compositionally induced convection in the OC, whereby denser (more Fe rich) crystals freeze out of the less dense (more alloyed) OC liquid. The crystals sink towards the IC, thereby inducing liquid motions that sustain the magnetic field. If Jephcoat and Olson's results are substantiated, the compositional (and hence density) difference between the IC and OC is less than previously thought and insufficient by itself to drive the dynamo. In this case the dynamo would have to be driven by thermal convection resulting from heat loss at the CMB. This would require a substantial temperature drop between the OC and mantle and hence a high heat flux into the D'' layer above the CMB. This would help in the formation of mantle plumes and the possible motion of the crustal plates at the Earth's surface. These questions will be further discussed in Chapter 6.

4

The thermal history of the Earth

4.1 INTRODUCTION

In everyday life we are accustomed to three forms of heat transport – conduction, convection and radiation. In the deep interior of the Earth, radiation is unlikely to be of any importance (although, as we shall see later, it may enhance the thermal conductivity). In the past, the thermal regime of the Earth has been assumed to be controlled by conduction alone. This has had two unfortunate consequences. The first is that the Earth has so much thermal inertia that the time τ required to reach equilibrium from some initial temperature distribution is longer than the age of the Earth. It can be shown that $\tau \simeq R^2/k$ where R is the radius of the Earth and k the thermal conductivity. For the Earth, $\tau \simeq 10^{18}$s. If one wished to boil some water to make a cup of coffee using only the heat flow from the Earth through an area equal to the cup's base, one would have to wait 15 years for the coffee! The second problem is the unknown initial conditions, which depend on the origin of the solar system. If convection occurs, these difficulties can be largely overcome – τ can be considerably reduced and, if the viscosity of the deep interior is essentially constant, as maintained by Tozer (1972) (and discussed later), then the physical properties of the interior have only a 'fading memory' of the initial conditions of the formation of the Earth and are not dependent on any particular cosmological theory. To obtain more specific ideas on the thermal regime of the Earth, we need to have estimates of the melting point and adiabatic temperature gradients and these will be discussed in sections 4.3 and 4.4 The importance of these gradients on the structure of the core is shown in section 4.2.

4.2 THE EARTH'S INNER CORE (IC)

A puzzling problem is how the Earth could have evolved into its present structure with a solid mantle, solid IC and liquid OC. A

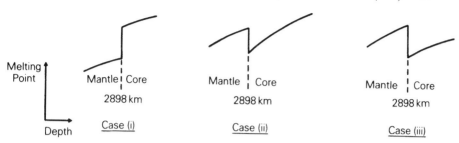

Figure 4.1 Possible forms of the melting point depth curve in the neighbourhood of the CMB.

possible explanation was put forward by Jacobs in 1953. If the transition from the IC to the OC is a transition from the solid to the liquid form of a single material, then the boundary must be at the melting point.

Consider the simplest type of model first – a core of pure iron and a mantle of silicates. At the boundary between the mantle and core (the CMB) there will be a discontinuity in the melting-point-depth curve, although the actual temperature must be continuous across the boundary. The form of this discontinuity could, mathematically, be any of the three cases shown in Figure 4.1. Case 1, in which the melting-point curve in the core is always above that in the mantle, is impossible; for the actual temperature curve must lie below the melting-point curve in the mantle, above it in the core, and yet be continuous across the boundary. Cases 2 and 3 are both possible. In case 3, the melting-point curve in the core never rises above the value of the melting point in the mantle at the CMB, while in case 2 it exceeds this value for part of the core. Considering first case 2, the melting-point curve will be of the general shape shown in Figure 4.2.

If the Earth cooled from a molten state, the temperature gradient would be essentially adiabatic, there being strong convection currents and rapid cooling at the surface. Solidification would commence at that depth at which the curve representing the adiabatic temperature first intersected the curve representing the melting-point temperature. Solidification would thus begin at the centre of the Earth. A solid IC would then continue to grow until a curve representing the adiabatic temperature intersected the melting-point curve twice, once at A, the boundary between the core and mantle, and again at B, as shown in Figure 4.2. As the Earth cooled still further, the mantle would begin to solidify from the bottom upwards. The liquid layer between A and B would thus be trapped. The mantle would cool at a relatively rapid rate, leaving this liquid

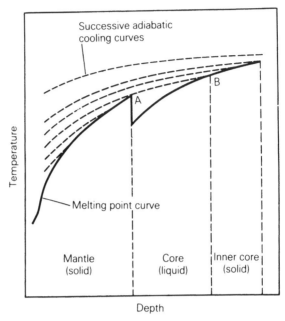

Figure 4.2 Melting point curve and successive adiabats in the Earth's interior (after Jacobs, 1953).

layer essentially at its original temperature, insulated above by a rapidly thickening shell of silicates and below by the already solid (iron) IC.

In the above discussion, no specific values of the temperatures are postulated and the behaviour of the adiabatic and melting-point curves need not be known exactly. If they vary qualitatively as shown, the above argument gives a physical explanation for the existence of a solid IC. It follows by similar reasoning that if the melting-point-depth curve in the neighbourhood of the CMB is as shown in case 3 of Figure 4.1, then as the Earth cooled from a molten state, the entire core would be left liquid. The physical state of the IC thus depends on the magnitude of the discontinuity in the melting-point curve at the CMB. As discussed in Chapter 3, the core is not pure iron but contains some light alloying element(s). As the core cooled, solidification would still begin at the centre of the Earth when the temperature first intersected the liquidus and a solid IC would grow as outlined above.

4.3 MELTING-POINT DEPTH CURVES

Both theoretical and experimental methods have been used to estimate the melting temperature of materials in the deep interior of the Earth. The fundamental relationship between the melting-point T_m, and pressure p is the Clausius-Clapeyron equation

$$\frac{dT_m}{dp} = \frac{\Delta V}{\Delta S} = T_m \frac{\Delta V}{L} \tag{4.1}$$

where ΔV and ΔS are the volume and entropy increments on melting and L the latent heat of melting. However this equation has not been used directly to compute melting temperatures in the deep interior of the Earth because of the unknown pressure dependence of ΔV and L. As a result, a number of semi-empirical relationships have been suggested. One that has been used extensively is Lindemann's law (Lindemann was Churchill's chief scientific adviser during the Second World War and was later made Lord Cherwell). Lindemann's law is essentially an extension of Debye's theory of lattice vibrations — melting occurs when the amplitudes of atomic vibrations become so large that atoms are forever colliding with one another. One short-coming of the theory is that it takes no account of the free energy of the molten state. At melting there are two phases in equilibrium (one a solid and one a liquid) whose energy densities may be substantially different. The Lindemann law can be expressed in terms of the Grüneisen parameter γ

$$\frac{1}{T_m} \frac{dT_m}{dp} = 2\left(\gamma - \frac{1}{3}\right) \bigg/ k \tag{4.2}$$

where k is the incompressibility along the melting curve. Grüneisen's parameter γ is a non-dimensional number defined as

$$\gamma = \frac{\alpha k_s}{\rho c_p} \tag{4.3}$$

where α is the volume coefficient of expansion, k_s the adiabatic incompressibility (equation (2.9)), and c_p the specific heat at constant pressure. One advantage of using γ is that its numerical value is close to unity for most condensed substances, and is thus more nearly constant over the range of physical conditions in the Earth's deep interior than are its component parameters α, k_s, ρ and c_p. Stacey and Irvine (1977) obtained a melting equation similar to equation (4.2) based purely on thermodynamic arguments. They used the Clausius-Clapeyron equation (4.1) together with the assumption that melting is only a minor perturbation of the crystal structure, atomic bonds being merely stretched or compressed.

Poirier (1986) used a dislocation theory of melting to estimate the melting temperature of Fe at the inner core boundary (ICB). The essential idea is that solids accumulate an increasing number of dislocations in their lattice structure as they approach their melting point. The greater the density of dislocations, the less the sense of order within a lattice and the more like a liquid it appears on a microscopic scale – in effect a liquid is merely a solid saturated with dislocations. Poirier used the dislocation theory developed by Ninomiya (1978) which gives the volume and entropy of melting and the slope of the melting curve as functions of the elastic moduli and Grüneisen's parameter, which may be obtained from seismology. Poirier obtained a melting temperature of Fe at the ICB of 6000 ± 300 K with a preferred value of 6150 ± 100 K. This is in fair agreement with estimates based on shock wave experiments and, rather surprisingly, with extrapolation based on Lindemann's law. Allowing for the depression in the melting point due to the addition of lighter elements in the OC, Poirier estimated the temperature at the ICB to be ~5000 K. He also calculated the density change on melting to be ~0.12 g cm^{-3}, which is about five times smaller than that at the ICB for the PREM model and more than an order of magnitude less than that estimated by Souriau and Souriau (1989) from amplitude ratios of the seismic waves PKiKP and PcP (section 2.7). This is important when we consider 'compositionally driven' convection in the OC as the mechanism which drives the geodynamo responsible for the Earth's magnetic field (section 5.3). It must also not be forgotten that iron exists in four crystalline forms (α, γ, δ and ε) so that the melting curve defines a condition of equilibrium between the liquid phase and one of the solid phases. Only at low pressures is the α (bcc) phase in equilibrium with the melt. There has been much discussion as to which phase γ (fcc) or ε (hcp) exists in the OC and IC.

Because of the importance in geophysics of the equation of state and phase diagram of iron at high pressures and temperatures, a workshop on the physics of iron was held in June 1989 at Los Alamos. The proceedings of this workshop have now been published (*J. Geophys. Res.* v. 95, no. B.13, pp. 21, 689–21, 776, 1990). One of the main reasons for holding the workshop was to discuss the conflicting estimates of the temperature of the liquidus of iron at high pressures.

Williams *et al.* (1987) measured the melting temperatures of γ Fe with a laser-heated diamond-anvil cell to pressures in excess of 100 GPa and with shock loading up to pressures of ~250 GPa (Figure 4.3). The melting point of γ Fe at the CMB (136 GPa) is 4800 ± 200 K. The authors estimate that the effect of a lighter, alloying element in the OC would be to lower the melting point of pure Fe by 1000 K. They also extrapolated their data to pressures of 360

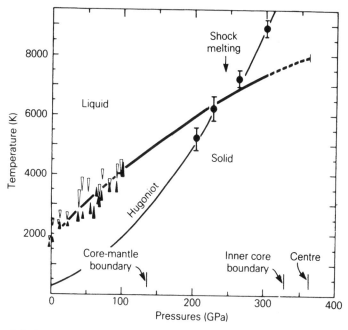

Figure 4.3 Summary of both static and dynamic determinations of the melting temperature of iron at high pressures. Circles represent measurements of Hugoniot temperatures in shock experiments and triangles the bounds obtained from static experiments. Solid triangles represent the highest temperature measurements on solid iron samples at a given pressure, and open triangles include the lowest temperature of liquid samples. The bold curve shows the estimate of the melting curve of iron (after Williams *et al.*, 1987).

GPa, the pressure at the centre of the Earth. Here the ε phase is the expected phase of iron. They estimate the melting temperature of ε Fe at the ICB (330 GPa) to be 7600 ± 500 K with an estimated liquidus temperature of 6600 K for the iron alloy in the OC. These temperatures are about 1500 K higher than the values obtained by Poirier. Boehler (1986, 1990), on the other hand, obtained much lower melting temperatures – a difference of more than 1000 K at 100 GPa compared to the values of Williams *et al.* Boehler suggested that there may be a new high pressure structural phase of iron – a high pressure bcc phase as well as the lower pressure bcc phase (ordinary α iron).

A discussion on the melting curve of iron has been given by Williams *et al.* (1991). They suggest that the difference between their estimates of the melting point and those of Boehler may be due to the measurements of the temperature gradient across the sample in the diamond anvil cell, Joule (resistance) heating (the so called 'wire heating') used by Boehler (1986) giving significant underestimates. Williams *et al.* (1991) believe that the best estimates of the high pressure melting curve of iron, derived from laser heated diamond anvil cell and shock wave measurements are their values of 4800 ($\pm$200) K at 133 GPa and 6700 ($\pm$400) K at 243 GPa.

Svendsen *et al.* (1989) have carried out liquid–state and solid–state model fits to the melting data for Fe, FeS and FeO which have a bearing on what is the light component in the OC (section 3.4). The temperature along the best fit Fe liquidus is 5000 K at 136 GPa and 7250 K at 330 GPa in agreement with the values of Williams *et al.* A comparison of data and solid-state model calculations indicate that FeS and FeO melt at 4610 and 5900 K respectively at the CMB and 6150 and 8950 K respectively at the ICB.

Knittle and Jeanloz (1987) have shown that laser-heated $Mg_{0.9}$ $Fe_{0.1} SiO_3$ remains in a perovskite-like structure to nearly the pressures found in the lower mantle and suggested that this mineral may be the most abundant in the Earth. They later (1989) determined its melting temperature to be 3800 ($\pm$300) K at a pressure of 96 GPa, and, by extrapolation to 136 GPa, a value of 4500 ($\pm$500) K at the CMB. These values provide an upper limit to the geotherm though the solid mantle, and are compatible with estimates of the temperature in the core being high. Temperatures at the CMB can thus be quite high without silicate perovskite melting. Some recent experiments, however, suggest lower temperatures. Stixrude and Bukowinski (1990) predict that the melting temperature of perovskite at the base of the mantle to be 3750 K.

4.4 ADIABATIC TEMPERATURES

The increase in temperature dT for a reversible adiabatic increase of pressure dp is given by

$$dT = \frac{\alpha T}{\rho c_p} dp \qquad (4.4)$$

where α is the volume coefficient of thermal expansion, ρ the density and c_p the specific heat at constant pressure. Assuming hydrostatic equilibrium, the variation of pressure with depth z is

$$\frac{dp}{dz} = g\rho \qquad (4.5)$$

Hence the adiabatic temperature gradient is given by

$$\frac{dT}{dz} = \frac{g\alpha T}{c_p} \qquad (4.6)$$

Since values of $g = g(z)$ are sufficiently well known, it is possible to integrate equation (4.6) for different values of α/c_p, assuming that the adiabatic and melting temperatures are the same at the ICB. Adiabatic and melting temperatures can thus be compared throughout the OC. It is obvious, however, from equation (4.6) that the adiabatic temperature gradient is critically dependent on the value of α/c_p and its possible radial dependence. The adiabatic gradient may also be written in terms of Grüneisen's parameter γ defined in equation (4.3). Thus

$$\frac{dT}{dz} = \frac{gT\rho\gamma}{k_s} = \frac{gT\gamma}{\phi} \tag{4.7}$$

were $\phi = k_{s/\rho} = V_p^2 - \frac{4}{3}V_s^2$ (equation 2.10) and is known from seismic data.

The adiabatic temperature gradient is quite sensitive to the assumed value of γ for iron at high pressures – in fact for a liquid, γ may be no more than a dimensionless combination of thermodynamic parameters and have no real connection with Grüneisen's parameter for a solid.

In estimating the adiabatic temperature in the convective parts of the Earth from equations (4.6) or (4.7), a constant of integration is needed – the value of T at some depth. This has been estimated from experimental results in a solid-state transition that can be identified as corresponding to a seismically determined transition. This in turn usually depends on assumptions about the chemical composition of the mantle and core. Most estimates have led to fairly low values of the temperature on the mantle side of the CMB. This would imply a large jump in temperature across the CMB if recent estimates of the melting temperatures of Fe, FeS and FeO are correct.

4.5 ANOTHER LOOK AT THE THERMAL REGIME OF THE EARTH'S CORE

Higgins and Kennedy (1971) found that the adiabatic temperature gradient in the OC was much steeper than the melting point gradient. This is just the opposite to what has usually been supposed, and, if true, the argument put forward by Jacobs (1953) for the formation of a solid IC and liquid OC would no longer be valid. Moreover since the OC is fluid, actual temperatures there must be above the melting point, so that, with Higgins and Kennedy's values, the actual temperature gradient in the OC would be much less than the adiabatic gradient. In this case the OC would be thermally stably stratified, thereby inhibiting radial convection.

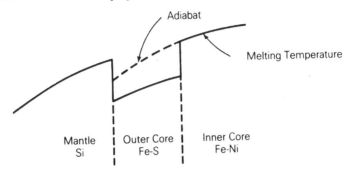

Figure 4.4 Possible melting point curve in the Earth and adiabat through the ICB (after Stacey, 1972).

This would pose problems for generating motions in the OC needed to power the geodynamo (section 5.3). It is interesting on looking back that the immediate reaction to Higgins and Kennedy's paper was to accept their findings at face value and try to think of ways to circumvent what they called the core paradox (Kennedy and Higgins, 1973). Thus Stacey (1972) suggested that if the light alloying element in the OC is sulphur (section 3.4.1), the IC consisting mainly of Fe–Ni, the presence of sulphur in the OC may so reduce its liquidus below that of Fe that the adiabat of Kennedy and Higgins through the ICB does not intersect it (Figure 4.4). The difficulties in estimating melting point and adiabatic temperature gradients have already been discussed – the melting point gradient of Higgins and Kennedy is particularly suspect. It is impossible to be dogmatic about these gradients, although it does seem now that the adiabat originating at the melting temperature at the ICB does lie entirely in the liquid phase, thereby removing the core paradox. In any case compositionally driven convection rather than thermal convection in the OC is now thought to be more likely to drive the geodynamo (section 5.3).

Since a light alloying element is required to explain the seismic data of the core, it is likely that the core has a melting interval rather than a well-defined melting temperature. If core temperatures are below the liquidus, solid iron would co-exist with an iron-rich melt solution. The solid iron could well settle into the centre of the core to form a solid IC, or be held in suspension by turbulent convection. In the former case the IC would grow with time. A general theory for the motion of an iron-alloy core containing a slurry has been developed by Loper and Roberts (1978). The state of the core depends on the magnitudes of three gradients of temperature with pressure – the gradients of the liquidus temperature

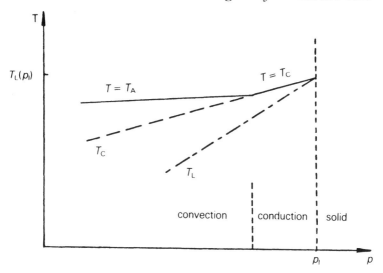

Figure 4.5 A schematic representation of the thermal regime $\dfrac{dT_A}{dp} <$ $\dfrac{dT_C}{dp} < \dfrac{dT_L}{dp}$. The fluid is both compositionally and thermally buoyant. The IC grows by direct freezing on its surface (after Loper, 1978).

T_L, the adiabatic temperature T_A, and the conduction temperature T_C. Of the six possible relative magnitudes, the cases

$$\frac{dT_L}{dp} < \frac{dT_A}{dp} < \frac{dT_C}{dp} \quad \text{and} \quad \frac{dT_L}{dp} < \frac{dT_C}{dp} < \frac{dT_A}{dp} \qquad (4.8)$$

are excluded, since they would imply a solid OC. The other four possibiities have been considered by Loper (1978).
    The case

$$\frac{dT_A}{dp} < \frac{dT_C}{dp} < \frac{dT_L}{dp} \qquad (4.9)$$

is shown in Figure 4.5. The solid Fe freezes directly on to the IC and the latent heat released there is removed in a thin conductive layer in which the temperature remains above the liquidus. The fluid is thus buoyant both compositionally and thermally.
    The case

$$\frac{dT_A}{dp} < \frac{dT_L}{dp} < \frac{dT_C}{dp} \qquad (4.10)$$

is interesting and is shown in Figure 4.6. The solid cannot freeze directly on to the IC because there is no conductive layer to remove

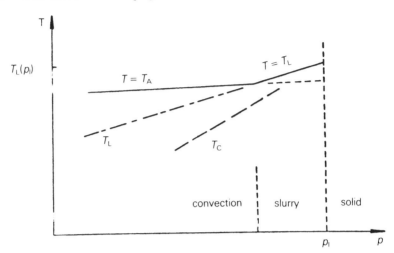

Figure 4.6 A schematic representation of the thermal regime $\dfrac{dT_A}{dp} <$ $\dfrac{dT_L}{dp} < \dfrac{dT_C}{dp}$. The slurry layer must form at the bottom of the OC. The IC grows by sedimentation of solid particles (after Loper, 1978).

the latent heat. Loper and Roberts (1981, 1983) have examined this case in more detail and shown that a slurry of solid particles would form suspended in the liquid phase above the ICB. The latent heat released by the freezing of the particles raises the temperature from the adiabat to the liquidus. The particles would freeze into a dendritic layer above the IC – such dendrites are tree-like structures which are easily broken. Loper and Roberts suggest that the slurry is thus more likely to be a mushy zone of dendrites between the liquid OC and the solid IC. It is just possible that the entire IC could be a mixture of liquid and solid – seismic evidence (the attenuation of seismic waves in the IC) has given some hints that the IC may be in a partially molten state. Finally there is the possibility that the solid IC grows by pressure-induced freezing as the proto-Earth grows by accretion, i.e., the cause of the freezing of the IC is not a decrease in temperature, but an increase in the central pressure as the mass of the Earth grows.

4.6 THE THERMAL HISTORY OF THE EARTH

The first calculations of the thermal history of the Earth were based on the theory of heat conduction alone. Two main processes account for heat conduction in solids. Below 1000 K energy transfer is mainly *via* thermo-elastic waves i.e., phonon or lattice conduction. The lattice conductivity of most materials decreases with increasing temperature, but increases with increasing pressure. For the Earth, pressure and temperature effects on lattice conductivity seem to counterbalance one another. Above 1000 K radiative heat transfer or photon conduction begins to dominate. At one time it was thought that radiative heat transfer at high temperatures would be important in the Earth. Spectroscopic investigations have now shown that mantle materials are too opaque for radiative transfer to be effective. Large scale convection is extremely efficient in transporting heat and such convective heat transfer will dominate thermal lattice and radiative heat conduction even for small velocities.

Some of the problems in treating conduction as the only mode of heat transfer in the Earth were discussed briefly in section 4.1. It was largely through the work of Tozer (e.g., Tozer 1977) that the importance of convection was really appreciated. Tozer showed that it is possible to estimate mean temperatures without knowing the exact details of the velocity distribution. Temperatures below a depth of about 800 km are controlled by the dependence of viscosity on temperature and for all plausible relationships, the temperature is always that which gives a viscosity $\sim 10^{20}-10^{21}$ poise. It is easy to see qualitatively how this could lead to a quasi-static regulated state. If the general temperature inside a planet were initially high, the associated low viscosity would lead to rapid convective loss of heat, and a fall of temperature that would slow the rate of cooling by increasing the viscosity. Alternatively, if the temperature were initially low and convective motion inhibited by very high viscosity, the interior would warm up to quite high temperatures because of the extreme slowness of the heat conduction process. Thus the idea that any convection theory must be very imprecise because of uncertainty in the viscosity-temperature relationship is not true; rather the viscosity itself is constrained to lie within very narrow limits. Tozer showed that, for a very large range of physically plausible values of material parameters, the temperature distribution in bodies larger than about 800 km in radius is very different from that predicted by conduction theory. In any body, the average temperature rises with depth according to the conduction solution, until either the centre of the body is reached or the kinematic viscosity has fallen to a value $\sim 10^{20}$ cm^2 s^{-1}; whichever is reached first. Once a body has become large enough for its central viscosity to be incapable of stabilizing a state of rest (i.e., conduction) solution, the steady central temperature becomes comparatively independent of the radius and surprisingly quite low (Figure 4.7).

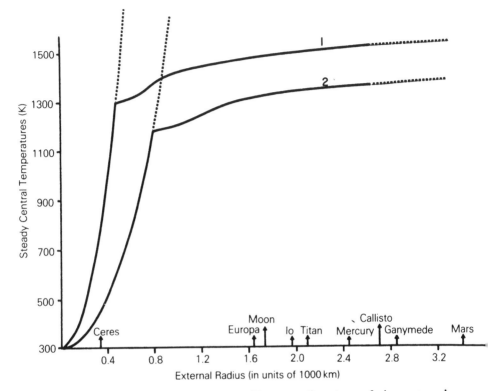

Figure 4.7 The central temperature T_C as a function of the external radius R for two values of the heat source density H. Curve 1, $H = 1.6 \times 10^{-14}$ cal. cm^{-3}s^{-1}, Curve 2, $H = 5 \times 10^{-15}$ cal. cm^{-3}s^{-1}. On the left the steeply rising curves are steady state of rest solutions – unstable where dotted. Note the relative independence of T_C on R when $R > 800$ km and the surprisingly low values of T_C (after Tozer, 1972).

As already mentioned it is now recognized that convection is the dominant mechanism of heat transport in the interior of the Earth. This would appear to make the heat-transport problem more difficult since the equations governing thermal convection are much more complicated than those governing conduction. Apart from the practical difficulties of solving the equations, the physical significance of any solution is limited by our lack of understanding of the thermodynamic and rheological parameters.

Apart from the impossibility of adequate planetary sampling, a fundamental barrier to an understanding of planetary science through laboratory experiments is knowing whether the incomplete information obtained in the laboratory is applicable to the same material in a body of planetary proportions. Because of its size, the evolution of a terrestrial planet is controlled by the rheological behaviour of its material under conditions quite different from any laboratory experiment. However if only the thermal history is desired with no interest in the details of the convective pattern, 'parameterized' convection models have proved useful. In this approach, the gross energy balance is considered without requiring an explicit solution for the dynamics. Instead of solving the complete set of equations which govern convection, parameterized convection uses a relationship between the Nusselt number Nu and the Rayleigh number Ra. The Rayleigh number is a dimensionless ratio of the buoyancy forces driving convection to the dissipative effects opposing convection. Convection sets in at a critical value of the Rayleigh number Ra_{cr}. The Nusselt number is a measure of the efficiency of convection in transporting heat; it is the ratio of the total heat-flux to the heat-flux without convection. Without convection, Nu = 1 whilst for rigorous convection Nu $\gg$ 1. Nu depends only on Ra and a relationship of the form

$$Nu = b \; Ra^{\beta}$$

or
$$NU = \left(\frac{Ra}{Ra_{cv}}\right)^{\beta} \tag{4.11}$$

where b and β are constants, has been found from experimental, theoretical and numerical studies. β lies between $\frac{1}{4}$ and $\frac{1}{3}$. The success of such a simple relationship is essentially due to the way in which convection works. Temperature profiles in fluids undergoing vigorous convection consist of thin thermal boundary layers with steep temperature gradients bounding interior regions in which heat advection is dominant and the temperature distribution is nearly isentropic. In the Earth, the lithosphere is the surface thermal boundary layer, and the D'' layer includes the basal thermal boundary layer, with the rest of the mantle being essentially isentropic. About ten times as much heat is being transported to the surface by convection as compared with conduction in the mantles of the Earth, Venus and Mars. Convective heat transport in the Moon may be several times the conductive heat transport, whilst in Mercury convective heat transport, if it is occurring at all, should be at best comparable to conductive heat flow.

Figure 4.8 shows a summary of experimental constraints on the present horizontally averaged temperature as a function of depth in the crust and mantle (the geotherm). Most of the temperature change occurs in thermal boundary layers in which vertical heat transfer is dominantly by conduction. Throughout most of the mantle advection is the primary mode of heat transfer, and the

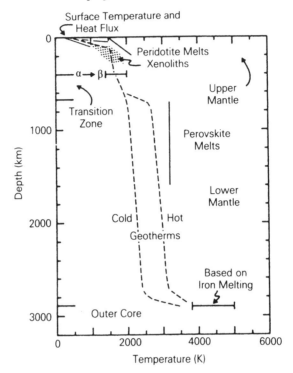

Figure 4.8 Summary of experimental constraints on the present, horizontally averaged temperature as a function of depth (the geotherm) through the crust and mantle. Two possible geotherms are shown for the lower mantle (hot and cold dashed lines) reflecting greater uncertainties in the average temperature at these depths. The basic form of the geotherm consists of adiabatic regions in which the temperature only increases slightly with depth and of thermal boundary layers in which large increases in temperature occur over a few hundred kilometres depth (after Jeanloz and Morris, 1986).

average temperature changes little with depth ($<0.5\,\mathrm{K/km}$). Horizontal temperature variations, however, are greatest in the central regions of the mantle being comparable to the temperature differences across the boundary layers.

The effect of temperature on the rheological properties of the mantle can be illustrated in terms of the Maxwell relaxation time (Figure 4.9)

$$\tau_M = \frac{\mu}{\eta} \tag{4.12}$$

where μ is the elastic shear modulus and η the effective viscosity. The response to a given shear stress is elastic over time periods

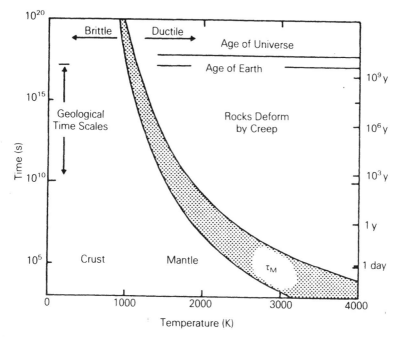

Figure 4.9 Maxwell relaxation time τ_M characterizing the ability of rocks to deform by solid state creep as a function of temperature. For a given shear stress, τ_M is the ratio of non-recoverable (creep) to recoverable (elastic) deformation that is achieved over a time period shown along the vertical axis. The curve is based on experimentally measured deformation properties and its width indicates the variability and uncertainty in the rheological properties of rock (after Jeanloz and Morris, 1986).

much shorter than τ_M and viscous over time periods much longer than τ_M. The relaxation time is dependent on temperature through the temperature dependence of viscosity

$$\eta = \eta_0 \exp\left(\frac{\phi}{kT}\right) \qquad (4.13)$$

where ϕ is the activation energy and k Boltzmann's constant. Since the bulk of the mantle is at temperatures of $\sim 2000\,\mathrm{K}$, Figure 4.9 shows it can behave like a viscous fluid even over short time periods.

5

The Earth's magnetic field

5.1 INTRODUCTION

At its strongest near the poles, the Earth's magnetic field is several hundred times weaker than that between the poles of a toy horseshoe magnet – being less than a gauss (Γ). Thus, in geomagnetism we are measuring extremely small magnetic fields and a more convenient unit is the gamma (γ), defined as $10^{-5}\,\Gamma$. Strictly speaking the unit of magnetic field strength is the oersted, the gauss being the unit of magnetic induction. The distinction is somewhat pedantic in geophysical applications since the permeability of air is virtually unity in cgs units. At the International Association of Geomagnetism and Aeronomy (IAGA) Scientific Assembly held in Kyoto in 1973, it was decided to adopt the SI system of units. In SI units $1\,\Gamma = 10^{-4}\,\text{Wb m}^{-2}$ (Weber/m^2) $= 10^{-4}\,\text{T}$ (tesla), Thus, $1\,\gamma = 10^{-9}\,\text{T} = 1\,\text{n T}$.

In a magnetic compass, the needle is weighted so that it will swing in a horizontal plane and its deviation from geographical north is called the declination, D. A non-magnetic needle which is balanced horizontally on a pivot, becomes inclined to the vertical when magnetized. Over most of the northern hemisphere the north-seeking end of the needle will dip downwards, the angle it makes with the horizontal being called the magnetic dip or inclination, I. The total intensity F, the declination D and the inclination I completely define the magnetic field at any point. The horizontal and vertical components of F are denoted by H and Z. H may be further resolved into two components X and Y, X being the component along the geographical meridian and Y the orthogonal component. Figure 5.1 illustrates these different magnetic elements. They are simply related to one another by the following equations.

$$H = F \cos I, \quad Z = F \sin I, \quad \tan I = Z/H \tag{5.1}$$

$$X = H \cos D, \quad Y = H \sin D, \quad \tan D = Y/X \tag{5.2}$$

$$F^2 = H^2 + Z^2 = X^2 + Y^2 + Z^2 \tag{5.3}$$

The variation of the magnetic field over the Earth's suface is best illustrated by isomagnetic charts, i.e., maps on which lines are

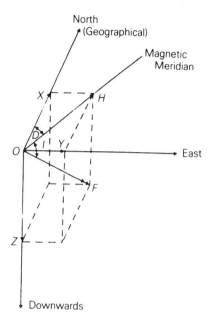

Figure 5.1

drawn through points at which a given magnetic element has the same value. Contours of equal intensity in any of the elements X, Y, Z, H or F are called isodynamics. Figures 5.2 and 5.3 are world maps showing contours of equal declination (isogonics) and equal inclination (isoclinics) for the year 1985. It is remarkable that a phenomenon (the Earth's magnetic field) whose origin, as we shall see later, lies within the Earth should show so little relation to the broad features of geography and geology. The isomagnetics cross from continents to oceans without disturbance and show no obvious relation to the great belts of folding or to the pattern of submarine ridges. In this respect the magnetic field is in striking contrast to the Earth's gravitational field and to the distribution of earthquake epicentres, both of which are closely related to the major features of the Earth's surface.

Not only do the intensity and direction of magnetization vary from place to place across the Earth, but they also show a time variation. There are two distinct types of temporal changes: transient fluctuations and long term secular changes. Transient variations produce no large or enduring changes in the Earth's field and arise from causes outside the Earth. Secular changes, on the other hand are due to causes within the Earth, and over a long period of time the net effect may be considerable. If successive annual mean values

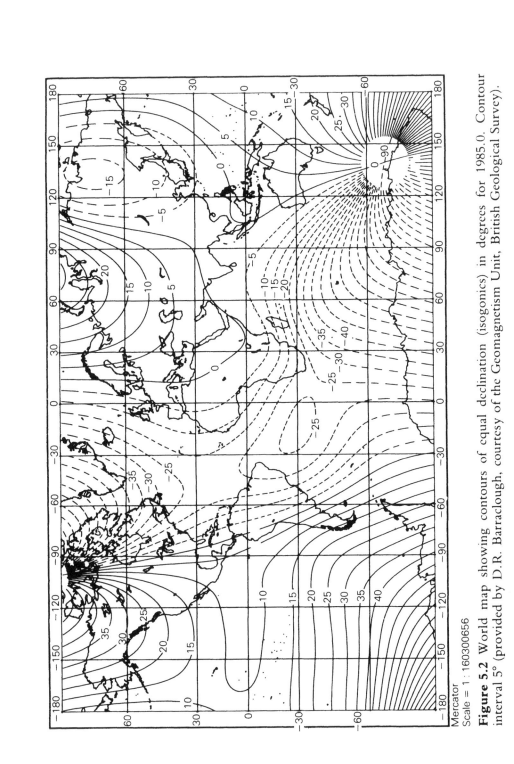

Mercator
Scale = 1 : 160300656

Figure 5.2 World map showing contours of equal declination (isogonics) in degrees for 1985.0. Contour interval 5° (provided by D.R. Barraclough, courtesy of the Geomagnetism Unit, British Geological Survey).

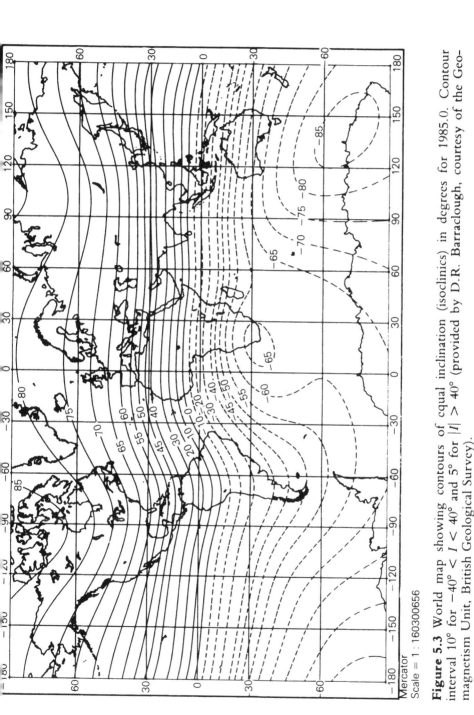

Mercator
Scale = 1 : 160300656

Figure 5.3 World map showing contours of equal inclination (isoclinics) in degrees for 1985.0. Contour interval 10° for −40° < I < 40° and 5° for |I| > 40° (provided by D.R. Barraclough, courtesy of the Geomagnetism Unit, British Geological Survey).

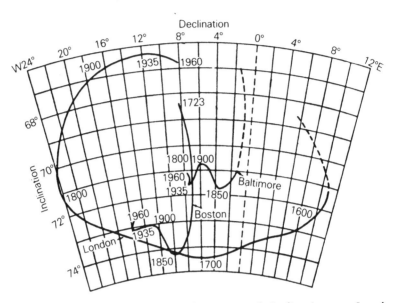

Figure 5.4 Secular change of declination and inclination at London, Boston and Baltimore (after Nelson *et al.* 1962).

of a magnetic element are obtained from a particular station, it is found that the changes are in the same sense over a long period of time, although the rate of change does not usually remain constant. Figure 5.4 shows the changes in declination and inclination at London, Boston and Baltimore. The declination at London was $11\frac{1}{2}°$E in 1580 and $24\frac{1}{4}°$W in 1819, a change of almost 36° in 240 y. Lines of equal secular change (isopors) in an element form sets of ovals centring on points of local maximum change (isoporic foci). Figure 5.5 shows the secular change in Z for the year 1987.5. If the pattern of secular change is compared at different epochs, it can be seen that the secular variation is a regional rather than a planetary phenomenon and that considerable changes can take place in the general distribution of isopors even within 20 years. The isoporic foci also drift westward at a fraction of a degree per year.

5.2 THE FIELD OF A UNIFORMLY MAGNETIZED SPHERE

Before 1600, William Gilbert had investigated the variation in direction of the magnetic force over the surface of a piece of the naturally magnetized mineral lode-stone which he had cut in the shape of a sphere. He found that the variation of the inclination

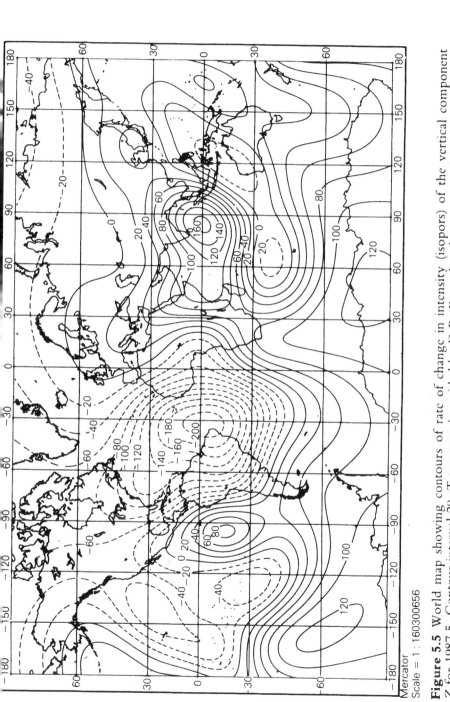

Mercator
Scale = 1 : 160300656

Figure 5.5 World map showing contours of rate of change in intensity (isopors) of the vertical component Z for 1987.5. Contour interval 20 nT per year (provided by D.R. Barraclough, courtesy of the Geomagnetism Unit, British Geological Survey).

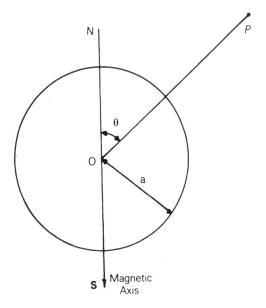

Figure 5.6

was in agreement with what was then known about the Earth's magnetic field, and came to the conclusion that the Earth behaved substantially as a uniformly magnetized sphere, its magnetic field being due to causes within the Earth and not from any external influence as was supposed at that time. Since 1600 the direction and intensity of the Earth's magnetic field have been measured at many widely scattered points over the Earth's surface, although no attempt was made to represent the field mathematically before 1839. In that year Gauss showed by a spherical harmonic analysis* that the field of a uniformly magnetized sphere which is the same as that of a dipole at its centre is an excellent first approximation to the Earth's magnetic field. Gauss further analysed the irregular part of the Earth's field, i.e., the difference between the actual observed field and that due to a uniformly magnetized sphere, and showed that both the regular and irregular components of the Earth's field are of internal origin.

* The variations of physical parameters over a sphere are most easily expressed using spherical harmonics. They are used in the treatment of gravity, seismology, heat transfer and magnetic fields. No details of such a representation will be attempted here. It must be stressed that a spherical harmonic analysis is just a mathematical convenience for describing global variations and does not necessarily have any physical significance.

Since the north-seeking end of a compass needle is attracted towards the northern regions of the Earth, those regions must have opposite polarity. Consider therefore the field of a uniformly magnetized sphere whose magnetic axis runs north-south, and let P be any external point distant r, from the centre O and θ the angle NOP, i.e., θ is the magnetic co-latitude (Figure 5.6).

If m is the magnetic moment of a geocentric dipole directed along the axis, the potential at P is

$$V = \frac{m}{4\pi} \frac{\cos\theta}{r^2} \tag{5.4}$$

The inward radial component of force corresponding to the magnetic component Z is given by

$$Z = -\mu_o\frac{\delta V}{\delta r} = \frac{\mu_o m}{2\pi} \frac{\cos\theta}{r^3} \tag{5.5}$$

and the component at right angles to OP in the direction of decreasing θ, corresponding to the magnetic component H, by

$$H = -\mu_o\frac{1}{r}\frac{\delta V}{\delta \theta} = \frac{\mu_o m}{4\pi} \frac{\sin\theta}{r^3} \tag{5.6}$$

where μ_o is the permeability of free space.

The inclination I is then given by

$$\tan I = \frac{Z}{H} = 2\cot\theta \tag{5.7}$$

and the magnitude of the total force F by

$$F = (H^2 + Z^2)^{1/2} = \frac{\mu_o m}{4\pi r^3}(1 + 3\cos^2\theta)^{1/2} \tag{5.8}$$

Thus intensity measurements are a function of latitude.

The geomagnetic poles, i.e., the points where the axis of the geocentric dipole which best approximates the Earth's field meets the surface of the Earth, are situated at approximately 79°N, 71°W and 79°S, 109°E. The geomagnetic axis is thus inclined at about 11° to the Earth's geographical axis.

The points on the Earth's surface where the field is vertical (i.e. the inclination is +90° and −90°) are called the magnetic or dipoles. They are not diametrically opposite and for epoch 1980 are at 77.3°N, 101°.8W and 65.6°S, 139.4°E. The geomagnetic and magnetic poles would coincide if the field at the Earth's surface was exactly that of a geocentric dipole.

Chevallier showed in 1925 that the remanent magnetization of several lava flows on Mount Etna were parallel to the Earth's magnetic field measured at nearby observatories at the time the flow erupted. If it is assumed that the geomagnetic field at the Earth's surface, averaged over several thousands of years (to eliminate the superficial secular variation), can be represented by a geocentric dipole with its axis along the axis of the Earth's rotation, it is possible by measuring the present direction of magnetization of a suite of rocks to deduce the position of the Earth's rotational axis relative to the location of the rocks at the time when they were laid down. The measured declination gives the azimuth of the land mass at the time the rock became magnetized, and the measured inclination gives the geographical latitude (equation 5.7). Unfortunately it is not possible to determine the original longitude. The inclination for a given latitude is the same for all longitudes, and the declination, although giving the direction of the pole, does not specify that longitude. The assumption that the Earth's field has corresponded on the average to that of an axial dipole situated at the centre is based on theoretical grounds and on measurements of rocks formed over the past few million years. If ancient pole positions are plotted in chronological order on the surface of the Earth for rocks from different areas, different 'polar wandering paths' are obtained, indicating that the continents have moved relative to one another.

The virtual geomagnetic pole (VGP) is defined as the pole of the dipolar field which gives the observed direction of magnetization at the site under consideration. It is calculated from any spot reading of the field direction, the word 'virtual' meaning that no implication about the position of an average dipole is being made. To compare data from different sampling sites at different latitudes it is convenient to calculate the equivalent dipole moment which would have produced the measured intensity at the calculated palaeolatitude (assuming a dipolar field) of the sample. Such a calculated dipole moment is called a virtual dipole moment (VDM).

5.3 THE ORIGIN OF THE EARTH'S MAGNETIC FIELD

There have been many suggestions for the origin of the Earth's magnetic field, although most of them have been shown to be inadequate. Consider first the possibility of permanent magnetization. The temperature gradient in the crust is approximately 30°C/km, so that at a depth of about 25 km a temperature of the order of the Curie point for iron is reached, and all ferromagnetic substances will have lost their magnetic properties at greater depths. Hence in order to account for the Earth's magnetic moment, an intensity of magnetization in the Earth's crust of about $600\,\mu T$ is necessary, which is impossible. The magnetization of rocks depends on the amount and nature of the iron oxide minerals contained in them, and for most rocks the intensity is less than $1\,\mu T$. Permanent magnetization also fails to account for other features of the Earth's magnetic field, such as the close proximity of the magnetic and geographical poles, the secular variation, and reversals of polarity.

Some workers have postulated changes in the fundamental laws of physics which would only become significant in rotating bodies of cosmic size. In this regard, Blackett in 1947 suggested that the dipole moment of a massive rotating body is proportional to its angular momentum. This was prompted by the discovery of Babcock in 1947 that the star 78 Virginis possessed a magnetic field. When Blackett first proposed his theory it appeared to predict correctly the relative magnetic moments of the sun and Earth in terms of their angular momenta. A revised value for the sun's magnetic moment later nullified this agreement. One consequence of Blackett's hypothesis is that a small magnetic field should be produced by a dense body rotating with the Earth. In a long and detailed paper, Blackett (1952) described the results of a 'negative experiment' in which a sensitive astatic magnetometer failed to detect a field of the order of magnitude predicted near dense bodies at rest in the laboratory. However, in developing an astatic magnetometer of greatly increased sensitivity, Blackett paved the way for the renewed activity in rock magnetism.

We are forced to the conclusion that electric currents flow in the Earth's interior and set up a magnetic field by induction. Electric currents may have been initiated by chemical irregularities which separated charges and set up a battery action generating weak currents. Palaeomagnetic measurements have shown that the Earth's main field has existed throughout geologic time and that its strength has never differed significantly from its present value. In a bounded, stationary, electrically conducting body, any system of electric currents will decay. For a sphere the size of the Earth this time is of the order of $100\,000\,y$. Since the age of the Earth is about $4500\,Ma$, the geomagnetic field cannot be a relic of the past, and a mechanism must be found for generating and maintaining electric currents to sustain the field.

A mechanism for generating and maintaining electric currents in the Earth's core is the familiar action of the dynamo. The dynamo theory of the Earth's magnetic field was due originally to Sir Joseph Larmor who in 1919 suggested that the magnetic field of the sun might be maintained by a mechanism analogous to that of a self-exciting dynamo. The pioneering work in dynamo theory was later carried out by Elsasser (1946a,b and 1947) and Bullard (1949a,b). The Earth's core is a good conductor of electricity and a fluid in which motions can take place, i.e., it permits both mechanical motion and the flow of electric current, and the interaction of these could generate a self-sustaining magnetic field. The development of the dynamo theory of the Earth's magnetic field has had to be based on theoretical models, because materials available for laboratory experiments are not sufficiently good conductors for the models to be of a reasonable size. If a bowl of mercury is heated from below, thermal convection will be set up, but no electric currents or magnetism will be detected in the bowl. Such a model experiment fails because electrical and mechanical processes do not scale down in the same way. An electric current in a bowl of mercury 30 cm in diameter would have a decay time of about one-hundredth of a second. The decay time, however, increases as the square of the diameter; an electric current in the Earth's core would persist for about 10 000 y before it decayed. This time is more than sufficient for the current and its associated magnetic field to be altered and amplified by motions in the fluid, even if very slow.

The energetics of the Earth's core will be discussed later but even if sufficient energy sources exist to maintain the field, there remains the critical problem of sign; it must also be shown that the inductive reaction to an initial field is regenerative. In an engineering dynamo, the coil has the symmetry of a clock face in which the two directions of rotation are not equivalent; it is this very feature which causes the current to flow in the coil in such a direction that it produces a field which reinforces the initial field. A sphere does not have this property; any asymmetry can exist only in the motions. This is the crux of the problem: whether asymmetry of motion is sufficient for dynamo action or whether asymmetry of structure is necessary as well.

More than fifty years ago, Cowling (1934) made a quantitative study of Larmor's suggestion that the magnetic fields of sunspots were the result of dynamo action due to motions of gas near the sun's surface. He examined steady, poloidal magnetic fields that were symmetric about an axis and proved that they could not be maintained by any self-sustaining dynamo process. Cowling's theorem does not rule out the dynamo mechanism, but it does show that it is difficult to find examples of dynamos that are mathematically simple – or even tractable. It is now recognized that departures from axial symmetry are an essential feature of the magnetic fields of both the sun and the Earth. Cowling's original form of the theorem has been extended by a number of workers (e.g., Hide and Palmer, 1982). It has been shown that no magnetic field that is symmetric about any axis can be sustained by dynamo action of a fluid. The fluid may be compressible and its properties may vary in space, and both magnetic field and fluid velocity may be time-dependent. Cowling's theorem does not preclude dynamo action by axially symmetric fluid motions, provided that the magnetic field does not have axial symmetry. The limitations imposed by Cowling's theorem led at one time to fears that dynamo action might in fact not be possible under any conditions. Fortunately this has not proved to be the case, although a number of other non-dynamo theorems that prohibit particular types of motion in a sphere from acting as dynamos have been obtained. It has been shown that in fact a great variety of different classes of fluid motion would lead to magnetic field generation – almost any sufficiently vigorous and sufficiently complicated fluid motion can generate a magnetic field.

The geomagnetic dynamo problem is a formidable mathematical problem and, despite many attempts and some limited success, as yet no completely satisfactory model has been produced. It is usually supposed that the fluid motion in the OC contains some differential rotation that winds up any dipole field present into a toroidal field – the field may be increased indefinitely by increasing the fluid motion. This does not solve the dynamo problem – to do so the dipole field must in some way be produced from the toroidal field, thus completing a cycle by which energy may be put into the field. This second stage of the cycle is much more difficult to account for. It should be pointed out that any toroidal field in the core would not be observed at the Earth's surface – it would be 'contained' within the core. It is generally believed that the toroidal field in the core is a good deal stronger than the poloidal field.

Most dynamo models use large scale, highly ordered fluid motions, i.e., motions in which the characteristic length of the velocity field is not much less than the radius of the Earth. In the early 1950s several attempts were made to produce models in which turbulent (i.e., random and small-scale) velocities might act as dynamos. The theory, which has been called mean field electrodynamics, has been developed independently by Moffatt (1970, 1978) in Britain and by Krause, Rädler and Steenbeck in Germany. No attempt will be made to discuss the theory here as an account of the German work has been given in a book by Krause and Rädler (1980).

Let us consider now the question of the driving force for the geodynamo. The energy could originate in a number of forms (gravitational, chemical, thermal) ultimately being converted into heat that flows out into the mantle. Palaeomagnetic measurements on rocks more than 3500 Ma old show that the Earth had a magnetic field at least that long ago. Moreover the maximum strength of the field has never varied by more than a factor of about 2. Thus the energy needed to drive the fluid motions in the Earth's core must have been available at more or less the same rate throughout most of the Earth's history. On the other hand, the energy source must not be too great or otherwise it would melt the mantle and produce more heat at the Earth's surface than is observed.

There are two main contenders for the energy source – thermal convection in the OC and the growth of the IC – freezing of material at the ICB would separate a heavy fraction (mainly Fe) leaving behind a lighter fraction in the OC that would be buoyant leading to compositionally driven convection. This was originally suggested by Braginskii (1963), following an earlier suggestion of Verhoogen (1961) that latent heat might be supplied by the cooling and freezing of liquid OC material to form the solid IC. There is a fundamental difference between a gravitationally powered dynamo and a heat driven dynamo. If the liquid OC is stirred by thermal convection, most of the heat will be convected away without any magnetic field being generated at all. Moreover, heat in the core can be dissipated not only by convection but by conduction as well, and conducted heat cannot contribute to dynamo action. It has been estimated that the efficiency of any heat driven dynamo is probably no more than 5–10%. On the other hand, the potential energy liberated from compositional convection can escape the system only by being converted first to kinetic energy and then to heat via some dissipative process.

Electrical heating is the major dissipative process in the core, because viscosity and the diffusion rate of the light material are so small. Thus a large part of the gravitational energy release will go into electrical heating, making the compositionally powered dynamo much more efficient than one driven by thermal convection. The persistence of the magnetic field of the Earth demands a constant energy source for at least the last 3500 Ma and this provides a constraint on the thermal evolution of the core. Gubbins *et al.* (1979) have used the equations of global energy and entropy balance to estimate the power source required for a specific magnetic field. The amount of power required depends on the exact nature of the source. Three possibilities were considered: radioactive heating in the core itself, loss of internal energy due to cooling and freezing of the OC to form the IC, and cooling of the whole core with consequent differentiation to form the IC with release of gravitational energy. The last of these includes all the sources except for radioactive heating, but the introduction of some radioactivity into the calculation would be a simple matter. For radioactive heating alone, 10^{13} W is required for the dynamo. This is just within the limits set by the observed surface heat flux (4 × 10^{13} W) and what some geochemists believe to be the heating due to K^{40}. Cooling itself cannot release enough heat to power the dynamo because the required cooling rate is so high that the IC would be a very recent feature of the Earth. The release of gravitational energy can provide a magnetic field of $1-2 \times 10^{-2}$ T, with the IC growing slowly to its present size over 4000 Ma, and a heat release of 2.5×10^{12} W. A lower heat flux is required because of the greater efficiency of conversion of gravitational energy into magnetic fields than heat.

5.4 THE EARTH'S MAGNETIC FIELD AT THE CORE-MANTLE BOUNDARY

If diffusion in the core can be neglected this implies for fluid dynamics that there is no conduction of heat and that the viscosity is zero, whilst for electrodynamics it means that the electrical resistivity is zero. Alfvén showed that if diffusion can be neglected, the magnetic field lines are 'frozen' into the fluid. The time scale of the secular variation (10–1000 y) is very much less than the diffusion time for the main magnetic field in the core ($\sim 10^4$ y). Thus as a first approximation, the core can be treated as a perfect conductor and Alfvén's theorem applies. The validity of the frozen flux approximation depends on the length and time scales of the disturbances causing the secular variation. Features in the magnetic field that are only a few hundred kilometres across have decay times that are comparable to the time scale of the secular variation and thus diffusion will have a significant effect on their evolution. Bloxham and Gubbins (1985, 1986) have shown that this is indeed the case. Bloxham (1986) has also invoked diffusion to explain changes in the flux patches in the South Atlantic region caused by radial upwelling of the field. It must be stressed that the relevant 'features' are the smallest actually present in the core, not those observed at the Earth's surface. Our knowledge of the core field is limited by the signal from magnetized crustal rocks which masks the very short wavelength signal from the core. Perfect conductivity requires that null flux curves (i.e., contours of zero radial magnetic field) retain their topology. This is a constraint that must be satisfied by the observations. The curves cannot appear or disappear, neither can they split in two or merge, although once two null flux curves were very close, very little diffusion would be needed to complete a merger.

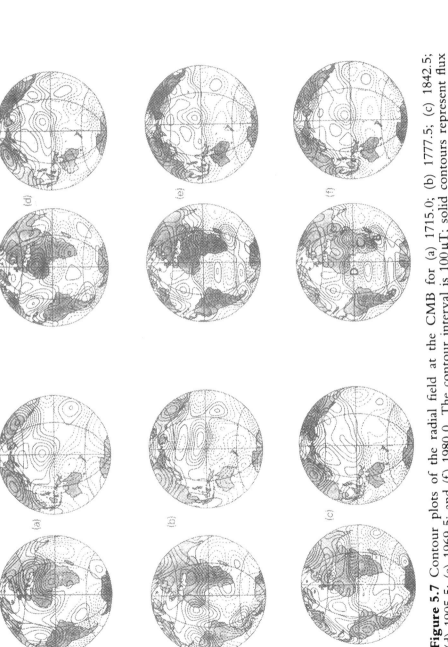

Figure 5.7 Contour plots of the radial field at the CMB for (a) 1715.0; (b) 1777.5; (c) 1842.5; (d) 1905.5; (e) 1969.5; and (f) 1980.0. The contour interval is 100 µT; solid contours represent flux into the core, broken contours flux out of the core. The bold contours represent zero radial field (after

Bullard suggested in 1958 that upwelling fluid in the OC crowds the field lines up against the CMB causing them to diffuse in and out of the boundary, creating an outward and an inward flux patch. Bloxham and Gubbins (1985) have called such a combination a core spot in analogy with sun spots. Their maps of the field at the CMB (cf. Figure 5.7f) confirm Bullard's suggestion. They found core spots beneath Africa and the Atlantic ocean, which, they suggest, are the cause of the observed weakening of the dipole component of the field.

A spherical harmonic analysis of the Earth's magnetic field observed at the surface of the Earth shows that a geocentric dipole is a very good first approximation. However in order to have a better understanding of the field, we need a picture of it, not at the surface of the Earth, but closer to its source, the OC. Measurements of the field at the surface can be extrapolated to the CMB by a spherical harmonic analysis. However there are complications since small scale surface features tend to blow up into large anomalies when projected down to the CMB – the higher harmonics, insignificant at the surface, assume much greater importance at depth.

Bloxham and Gubbins (1985) obtained models of the magnetic field at the CMB at selected epochs from 1715.0 to 1980.0 using in all cases the original magnetic field observations. The radial component of the magnetic field at the CMB is plotted for each of these models in Figure 5.7. The resolution of small-scale features is remarkable even in the 1715.0 field model. Bloxham and Gubbins (1985) identified a number of features:

(i) static flux bundles (permanent regions of intense flux observed under Arctic Canada, Siberia and Antarctica, the central Pacific Ocean and the Persian Gulf);

(ii) static zero-flux patches (permanent regions of very low flux observed at the North Pole, under Easter Island, in the northern Pacific Ocean, and in many models near the South Pole);

(iii) rapidly drifting flux spots (observed in the southern hemisphere from around 90°E, drifting westward towards South America with changes in intensity. Four such spots labelled A–D in the plot for 1980 in Fig. 5.7f can be traced through many epochs);

(iv) localized field oscillations (such as that under Indonesia).

This picture of the field at the CMB is very different from what has previously been inferred from maps of the surface field. Westward drift occurs only in certain well-defined regions of the

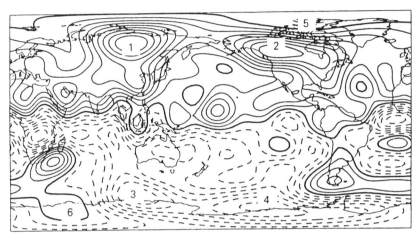

Figure 5.8 Map of the radial component of the magnetic field at the CMB for 1980. Contour interval is 100 µT; solid contours represent flux into the core, broken contours flux out of the core; bold contours represent zero radial field. The two main pairs of lobes (1,3) and (2,4) are indicated, as are the patches of low radial field (5 and 6) near the poles (after Gubbins and Bloxham, 1987).

core – in the northern Pacific Ocean there is slow eastward drift. It is evident that these models are not consistent with the frozen-flux approximation, so that the effects of magnetic diffusion must be taken into consideration in estimating fluid flow in the Earth's core.

Figure 5.8 shows a map of the radial component of the magnetic field at the CMB for 1980 (Gubbins and Bloxham, 1987). Some features are stationary. By far the most striking are the four flux concentrations symmetrically placed about the equator marked 1–4. With a typical westward drift rate of 0.2°/y, these features would have drifted over 50° in longitude during the period 1915–1980 covered by the analysis of Bloxham and Gubbins (1985). Such a large drift would have been detectable in their analysis and has almost certainly not occurred. Bloxham and Gubbins suggest that these four patches are the tops and bottoms of two columns of liquid which touch the IC and run parallel to the Earth's rotational axis. Fluid, spiralling down through the two columns, creates a dynamo process that concentrates flux within the columns. Such a model agrees with theoretical and experimental work by Busse (1970) who built a model consisting of two spheres – the smaller sphere rotating inside a larger sphere filled with water (Figure 5.9).

Many attempts have been made to determine the direction of the fluid flow in the Earth's core from observations of the magnetic field at the Earth's surface. Unfortunately such calculations have, in

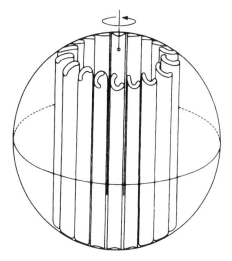

Figure 5.9 Convection rolls in a rapidly rotating sphere (after Busse, 1970).

general, no unique solution. The determination is not unique even with the frozen flux approximation. The ambiguity can be removed by imposing additional conditions on the velocity, e.g., time independence, geostrophy, density stratification allowing only toroidal motions. It is difficult to evaluate each of these additional constraints by its consistency with observations.

It is possible that convection of heat through the mantle could cause temperatures at the CMB to vary, thus influencing fluid flow in the OC and the magnetic field. Bloxham and Gubbins (1987) have compared their maps of the magnetic field at the CMB with temperature variations in the lower mantle as determined by Dziewonski (1984) and Hager *et al.* (1985) from variations in the seismic velocities and geoid anomalies. They noticed that three of the four flux patches that contribute strongly to the dipole field lie under cold regions. Moreover core spots are seen under South Africa where the mantle is hot. There is also a low flux feature in the Pacific Ocean below a hot region of the mantle. Bloxham and Gubbins suggest that a cold ring around the Pacific Ocean may be the reason for the lack of magnetic features there.

5.5 REVERSALS OF THE EARTH'S MAGNETIC FIELD

Apart from its spatial variation, the Earth's magnetic field also shows temporal changes ranging from variations on a timescale of seconds to secular variations on a timescale of hundreds of years and on an even longer timescale to complete reversals of polarity. The short period, transient variations are due to external (solar) influences and have no lasting effect on the Earth's main magnetic field, which, as we have seen, is of internal origin. They will not be discussed at all in this book. Variations in the time range $10-10^4$ y (the secular variation) have already been discussed. Reversals occur at widely varying intervals from about 30 000 y to more than 10 Ma, while variations in the frequency of reversals have time constants of the order of 50 Ma.

Although most rock-forming minerals are non-magnetic, all rocks show some magnetic properties due to the presence of various iron oxide and sulphide minerals making up only a few per cent of the rock. These minerals occur as small grains dispersed through the magnetically inert matrix provided by the more common silicate minerals that make up most rocks. When a rock forms it usually acquires a magnetization parallel to the ambient magnetic field referred to as a primary magnetization. This can give information about the direction and intensity of the magnetic field in which the rock formed. However, subsequent to formation, the primary magnetization may decay either partly or wholly and further components may be added by a number of processes. These subsequent magnetizations are called secondary magnetization. A major problem in palaeomagnetic investigations is to recognize and eliminate secondary components (O'Reilly, 1984).

The mechanism by which magnetism is acquired depends upon the mode of formation and subsequent history of the rocks as well as the characteristics of the magnetic minerals they contain. Rocks may become magnetized by a number of different natural processes, the most important of which are thermoremanent magnetization (TRM) and depositional remanent magnetization (DRM).

Igneous rocks cooling from above their Curie temperature in a magnetic field acquire a remanent magnetization, called thermoremanent magnetization (TRM). This magnetization is parallel to the applied field and for low field strengths is directly proportional to it. If a rock is cooled through various temperature intervals in the presence of a magnetic field, the TRM acquired in each interval is found to be independent of that acquired in each of the other intervals. The total magnetic moment is the sum of all the magnetic moments of the individual grains. The moment is not stabilized until the grain is cooled below the 'blocking temperature'. A piece of clay will normally contain minerals with blocking temperatures from room temperature to 680°C. Depositional remanent magnetization (DRM) is acquired when a grain falls through water in the presence of a magnetic field. Usually the DRM of most sediments is formed after the grains have come to rest on the bed by rotation in water-filled interstices. This post-depositional remanent magnetization (PDRM) is locked into the sediment by consolidation. This may take from a few days or weeks up to tens of thousands of years. The nature of the record thus varies considerably depending on the recording medium. A thin lava flow may acquire all its primary magnetization in less than a year whilst a sample of marine sediments may give some average value of the magnetic field over a time span of a few hundred to several thousand years, depending on the sedimentation rate.

Many rocks show a permanent magnetization that is approximately opposite to that of the present field. Reverse magnetization was first discovered in 1906 by Brunhes in a lava from the Massif Central mountain range in France; since then examples have been found in almost every part of the world. In 1966, Opdyke *et al.* found a polarity record in deep sea sediments going back 3.6 Ma in which the pattern of reversals was remarkably similar to that found in igneous rocks on land (Figure 5.10). Earlier there had been some controversy as to whether reversals of magnetization were the result of reversals of the Earth's field or whether there exist some physical or chemical processes whereby a material could acquire a magnetization opposite in direction to that of the ambient field. No two substances could be more different or have more different histories than the lavas of France and the sediments of the Pacific. The lavas were poured out, hot and molten, by volcanos and magnetized by cooling in the Earth's field (TRM); the ocean sediments accumulated grain by grain by slow sedimentation and by chemical deposition in the cold depths of the ocean (PDRM). If these two materials show the same pattern of reversals then it must be the result of an external influence working on both. Although it has been found that self-reversal does occur in some rocks, the evidence overwhelmingly supports field reversal as the cause of reversals of magnetization.

To prove rigorously that a reversed rock sample has become magnetized by a reversal of the Earth's field, it is necessary to show that it cannot have been reversed by any physico-chemical process. This is almost impossible to do since physical changes may have occurred since the initial magnetization or may occur during laboratory tests. More positive results can only come from the correlation of data from rocks of various types at different sites and by statistical analyses of the relationship between the polarity and other chemical and physical properties of the rock sample. There is no *a*

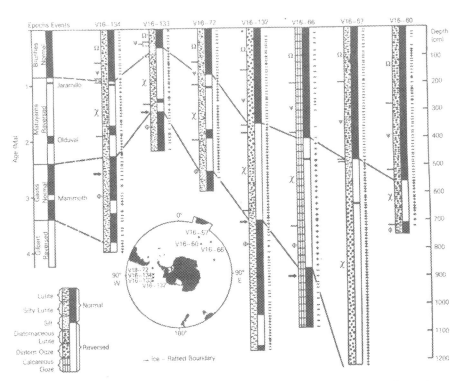

Figure 5.10 Correlation of magnetic stratigraphy in seven cores from the Antarctic. Greek letters denote faunal zones. Inset: source of cores (after Opdyke *et al.*, 1966).

priori reason why the Earth's field should have a particular polarity and there is no fundamental reason why its polarity should not change. The basic equations of electrodynamics and hydrodynamics in the Earth's core remain unchanged if the sign of the magnetic field is reversed although this is not true for the velocity field. Thus if a given velocity field will support a magnetic field, it will also support the reversed field and the same forces will drive it.

During a polarity change, the direction of the geomagnetic field swings through about 180°, the poles following widely different paths for different transitions. This could be interpreted as the result of a decrease in the main dipole field so that the observed field is dominated by the non-dipole component. Alternatively, the field could just tip over being still mainly controlled by the dipole term. This does not seem to be the case. Moreover the intensity of the field does not appear to vanish – the minimum value seems to be about 10–20% of the mean field intensity outside the reversal. The decrease in intensity during the first part of the transition is more likely to be due to a reorganization of the fluid motions in the core rather than by free decay of the field.

Four major normal and reversed sequences have been found during the past 3.5 Ma. These major groupings were originally called polarity epochs and named after people who have made significant contributions to geomagnetism. Superimposed on these polarity epochs are brief fluctuations in magnetic polarity with a duration that is an order of magnitude shorter. These were originally called polarity events and named after the localities where they were first recognized (Figure 5.11). The terms epoch and event have now been superseded by a system of magneto-stratigraphic polarity units (Harland *et al.*, 1990).

A polarity transition takes place so quickly (on a geological time-scale) that it is difficult to find rocks that have preserved a complete and accurate record. Good intensity estimates may be obtained for volcanic rocks but suffer from the fact that there is little chronological control. On the other hand, sedimentary rocks give reasonably good chronological control, but sedimentation rates are often too slow to allow detailed resolution of intermediate fields. It seems that during a reversal the intensity of the field first decreases by a factor of 3 or 4 for several thousand years while maintaining its direction. The magnetic vector then usually executes several swings of about 30°, before moving along an irregular path to the opposite polarity direction, the intensity still being reduced, rising to its normal value later. It is not certain whether the field is dipolar during a transition. Moreover there do not seem to be any precursors of a reversal or any indication later that a reversal has occurred.

In addition to polarity changes, the Earth's magnetic field has often departed for brief periods from its usual near-axial configuration, without establishing, and perhaps not even instantaneously approaching, a reversed direction (Figure 5.12). This type of behaviour has been called a geomagnetic excursion or aborted reversal. Geomagnetic excursions have been reported in lava flows of various ages in different parts of the world and from deep-sea and lake sediments. Excursions are generally observed to commence with a sudden and often fairly smooth movement of the pole towards equatorial latitudes. The pole may then return almost immediately, or it may cross the equator and move through latitudes in the opposite hemisphere before swinging back again to resume a near-axial position. Most geomagnetic excursions have been recorded in lake or deep-sea sediments and it is extremely difficult to assess the reliability of ages assigned to them. The

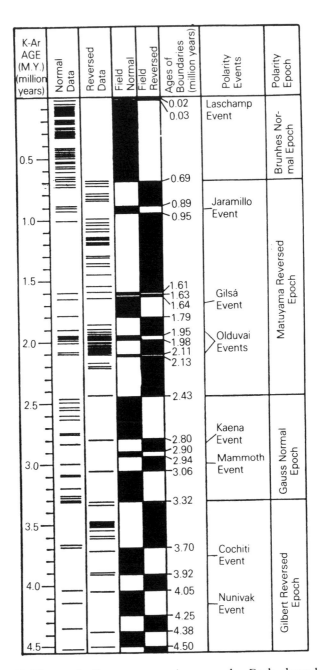

Figure 5.11 Timescale for geomagnetic reversals. Each short horizontal line shows the age as determined by potassium–argon dating and the magnetic polarity (normal or reversed) of one volcanic cooling unit. Normal polarity intervals are shown by the solid portions of the 'field normal' column and reversed polarity intervals by the solid portions of the 'field reversed' column. The duration of events is based in part on palaeomagnetic data from sediments and magnetic profiles (after Cox, 1969).

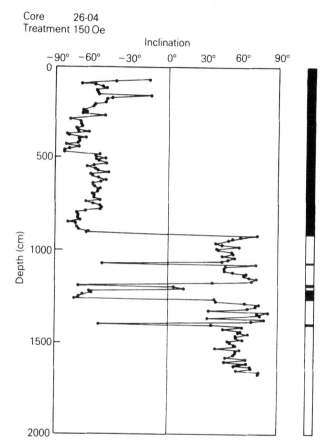

Figure 5.12 Inclination of short-period events seen in a deep-sea core from the Southern Ocean. Polarity log at right, clear is reversed (after Watkins, 1968).

horizons are usually dated by assuming uniform sedimentation rates between or beyond ^{14}C dated horizons.

Irving and Pullaiah (1976) have analysed the palaeomagnetic data over the last 600 Ma in order to find the proportion of normal to reversed polarities. Their results (Figure 5.13) clearly show polarity 'bias' which had been noted earlier by McElhinny (1971). For long intervals of geologic time it was reversed most of the time and during other intervals it was normal most of the time. Thus there is a marked difference between the Lower and Middle Devonian, which have roughly equal occurrences of normal and reversed data, and the Upper Devonian which is largely reversed and marks the beginning of a predominantly reversed era lasting until the close of

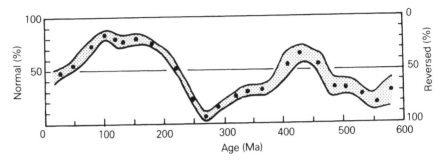

Figure 5.13 Polarity bias of the geomagnetic field during the Phanerozoic. Overlapping 50 Ma averages of polarity ratios as observed in palaeomagnetic results are shown together with the limits of the standard errors (after Irving and Pullaiah, 1976).

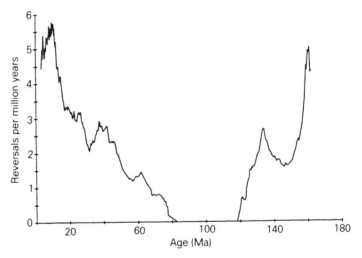

Figure 5.14 Estimated mean reversal rate from the present back to 165 Ma (after Merrill and McFadden, 1990).

the Palaeozoic. For the mid-Carboniferous to the Lower Permian, only a few results show normal polarity. Apart from the Lower Triassic, where frequent reversals have been observed in several formations, the Mesozoic has predominantly normal polarity (~75%); measurements in the Upper Triassic indicate that the field was normal as much as 83% of the time.

The frequency of reversals has also varied over geologic time (Figure 5.14). The rate of reversals appears to have decreased in an approximately linear manner with time from about 5 per Ma about 165 Ma ago until it ceased at about 107 Ma, resulting in what has been called the Cretaceous Normal polarity interval. At about 86 Ma ago, the process of reversals began again, the rate increasing approximately linearly with time. There is some indication that the rate of reversals reached a maximum of about 5 to 6 per Ma about 10 Ma ago and is again decreasing (McFadden and Merrill, 1984). Another 'quiet' interval (the Kiaman) has been observed within the upper Carboniferous and Permian (about 235–290 Ma ago) – for about 60 Ma the Earth's magnetic field appears to have been almost always reversed.

There have been a number of attempts to analyse statistically the sequence of reversals. If it is assumed that reversals are due to instabilities in the magneto hydrodynamic (MHD) processes in the core, a key question is the length of memory of such processes, 'length of memory' being used to describe the maximum time during which coherency persists in the pattern of core processes. The time constants of fluid motions in the Earth's core as deduced from the geomagnetic secular variation appear to range from $10–10^4$ y – similar time constants appear to be involved in the process of polarity switching, which is completed in about 4×10^3 y. If these time constants are characteristic of the longest time constants of MHD processes in the core, the implication is that the magnetic memory of the core is short relative to the mean length of polarity intervals, which varies from about 2×10^5 to about 10^6 y during the Cenozoic. If this is true, Cox (1968) showed that the sequence of reversals for the past 70 Ma is reasonably well satisfied by the familiar Poisson process of communication theory,

$$P(\tau) = \lambda \exp(-\lambda\tau), \qquad (5.9)$$

where the parameter λ characterizes the observed variations in the length τ of polarity intervals. However, the observed number of short polarity intervals is somewhat smaller than that predicted. With the discovery of previously undetected short events, the fit between the observed and predicted distributions has been improved. It is difficult to obtain evidence for brief periods of polarity change, since sediment accumulation in ocean basins may be too slow, and the polarity time scale is based on the age and polarity of terrestrial volcanic rocks. One of the problems in any statistical study of the sequence of reversals is that there are just two possibilities; normal or reversed. The discovery of a short reversed interval in a long period of normal polarity replaces this long period of normal polarity with two much shorter periods of normal polarity and a very short period of reversed polarity. In a non-stationary Poisson process, there is no memory of how long it has been since the previous event. Thus a reversal neither inhibits nor encourages future reversals, and long term changes in reversal rate must be caused by changes in the inherent rate of dynamical processes in the core.

We can well ask how rapidly can the Earth's magnetic field change? The maximum rate of the secular variation in recent times derived from a spherical harmonic analysis is ~150 nT/y. The highest resolution palaeomagnetic records give high rates of change of direction of ~0.5–2°/y – estimated from changes in direction of several tens of degrees during excursions and reversals. Mankinen *et al.* (1985) and Prévot *et al.* (1985) carried out a detailed investigation of the Miocene R → N polarity transition recorded in lava flows from Steens Mountain in South-east Oregon. Two or three episodes of extremely rapid field variations were found during the transition, larger by a factor of 30–40 than the above rates. On one occasion, the directional change was 58 ± 21°/y, while the intensity change was 6700 ± 2700 nT/y which is ~50 times larger than the maximum rate of change of the non-dipole field observed in historical records. In a later paper, Coe and Prévot (1989) found evidence of even greater changes in direction of high temperature remanence as a function of vertical position in one basalt flow. They considered a number of possible explanations and finally suggested that the anomalous remanence directions are the result of an 'impulsive change' in the transitional field during the reversal. If this is true, the short time required for the flow to cool by conduction and acquire its TRM implies rates of change of the magnetic field of at least 3° and 300 μT/day! Such rapid changes would imply fluid velocities near the CMB of the order of 1 km/h, about two orders of magnitude greater than that indicated by recent analyses of the secular variation. Moreoever mantle shielding should smooth out such rapid changes. It is easier to believe that there were complications, as yet undetected, in the palaeomagnetic record, such as a chemical remanent magnetization overprint, than to accept the data as a true record of the Earth's magnetic field.

Periods of rapid change have also been reported by Laj *et al.* (1988), although at least an order of magnitude slower than those suggested at Steens Mountain. They analysed records of directional and field intensity changes occurring during three sequential polarity reversals obtained from marine clays in the island of Zakinthos (Greece). Periods of very rapid change alternate with periods of near stationarity. However the records are quite different from those of four sequential reversals from marine clays from western Crete in the same geographical area, but separated in time by about 5 Ma. This shows that it is difficult to draw general conclusions from the study of single isolated records – different transition mechanisms may occur at different times.

The records from three different environments: the lavas from Steens Mountain; the Tatoosh intrusives from Washington; and the marine sediments from Greece, all support the contention that during reversal attempts the field can move rapidly as well as remain stationary and can undergo directional rebounds and unsuccessful attempts to complete the reversal process. There appear to be intermediate dynamo states (first noticed by Shaw, 1975) which act as a kind of springboard from which reversal attempts, both successful and unsuccessful, are made. Perhaps during rapid directional jumps, unusually high accelerations in the fluid flow in the OC drive the frozen flux into continuously changing patterns, and when the field is relatively quiet, diffusion may be important.

From the geomagnetic record alone, there is little hope of deciding whether the reversal sequence is chaotic (section 1.4). Systems that exhibit chaotic behaviour usually exhibit non-chaotic behaviour for different parameter values, and it is not easy to specify values appropriate to the Earth's dynamo. Again the lack of predictability associated with sensitivity to initial conditions means that one should not, and indeed cannot, account for every observed fluctuation. Moreover we do not know how long a period might occur without a reversal being a natural fluctuation rather than as something requiring a special explanation (as for example the Cretaceous Normal interval and the Kiaman reversed interval). Again we do not know to what extent trends in the geomagnetic data (e.g. variations in the average reversal frequency) reflect major changes in the conditions in the core, or are produced by only minor changes to which the statistics are sensitive, or indeed, might even occur without any causal change at all.

Prévot *et al.* (1990) have produced a new compilation of all Triassic and younger palaeointensity determinations. They used a selected data set in an attempt to minimize uncertainties inherent in intensity measurements. A major difficulty in interpreting palaeointensity data is removing the effects of palaeosecular variation. Their results are shown in Figure 5.15. It can be seen that from the Cretaceous-Tertiary boundary to Pleistocene times, the average virtual dipole moments (VDMs) are close to the present value for the past $10\,000$ y ($8.75 \times 10^{22}\,Am^2$), indicating a relatively constant field strength for the whole of the Cenozoic. On the other hand, the evidence indicates a low VDM close to $3 \times 10^{22}\,Am^2$ from the middle Jurassic to the Early Cretaceous. The rather sparse data are compatible with a near-present field intensity from Lower Triassic to Lower Jurassic times. Figure 5.15 also shows that the palaeointensity changes – at least during the period from the present to $160\,Ma$ ago for which the data are best documented – seem

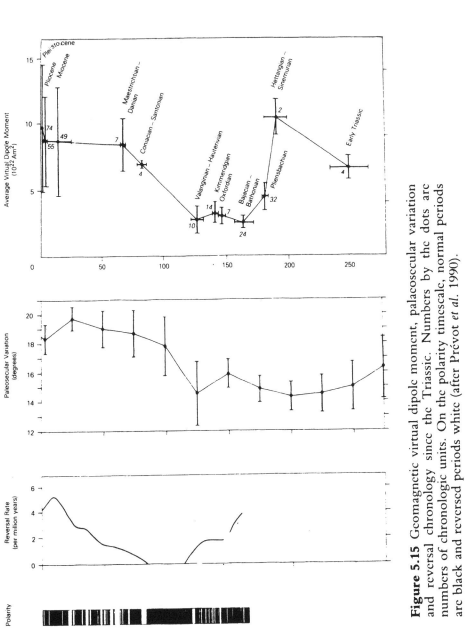

Figure 5.15 Geomagnetic virtual dipole moment, palaeosecular variation and reversal chronology since the Triassic. Numbers by the dots are numbers of chronologic units. On the polarity timescale, normal periods are black and reversed periods white (after Prévot *et al.* 1990).

to correlate with the changes with time of the amplitude of the directional palaeosecular variation.

Large long-term variations in the strength of the geomagnetic field must result from major changes in the operation of the geodynamo. The process causing the intensity variation is probably unrelated to the reversal process, during which the intensity decrease is accompanied by large directional changes and dominance by non-dipole components. Figure 5.15 suggests that most of the Mesozoic dipole low interval corresponds to a progressive decrease in reversal frequency. The end of the Mesozoic low probably lies somewhere in the long Cretaceous normal interval.

In conclusion, it must be emphasized that we do not at present fully understand the workings of the Earth's dynamo. We have already discussed the power source and reversals, but there are many unanswered questions. What decides the spatial structure of the field (a dipole tilted with respect to the rotational axis and a complex non-dipole field)? What is the cause of the secular variation and its time structure? Why does the field strength fluctuate over periods of the order of 10^4y, yet its total magnitude (apart from reversals) changes so little? Can one fit what we know about the geomagnetic field into a general scheme that explains the existence or non-existence of other planetary magnetic fields (Chapter 7), or are there several different possible mechanisms?

__6

The core-mantle boundary and geodynamics

6.1 INTRODUCTION

The core-mantle boundary (CMB) and the structure of the 200–300 km layer at the bottom of the mantle (the so-called D'' layer by Bullen are two of the most controversial problems in geophysics today. It is now believed that dynamic and chemical processes there strongly influence mantle convection, the formation of plumes, the secular variation of the Earth's magnetic field (and possibly reversals), long wave length gravitational variations and the chemical evolution of the Earth. In the past information on the CMB transition zone has come mainly from seismology. In recent years, geomagnetism, mineral physics and geodynamics have all added to our knowledge of the region which turns out to be even more complex than we thought. One of the main unresolved questions is whether there is a thermal and/or chemical boundary layer at the CMB.

Convection driven by lateral differences in temperature T and density ρ will lead to differences in the seismic velocities V_P and V_S – both V_P, V_S and ρ will decrease as T increases. Compositional variations, partial melting and anisotropy can also perturb V_P and V_S and changes in ρ due to these causes can also cause convection. By mapping V_P and V_S throughout the Earth we can infer changes in T and ρ that drive mantle convection. Measurements of seismic wave anisotropy can also yield information on the direction of flow in the mantle. One question that still remains unanswered is whether convection in the mantle is one or two stage, i.e., whether the entire mantle is freely convecting or whether there are two separate systems with a boundary at the seismic discontinuity at a depth of ~670 km (Figure 6.1). We shall return to this question later.

6.2 EVIDENCE FROM SEISMOLOGY

Many seismic models of the D'' layer have been proposed (see Young and Lay, 1987 for a good review). Some seismologists maintain that

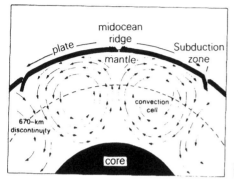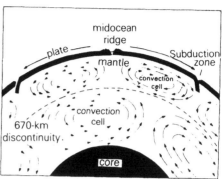

Figure 6.1 Schematic cross-section of convection in the mantle showing two possible models – whole mantle convection and two stage convection (after Dziewonski and Anderson, 1984).

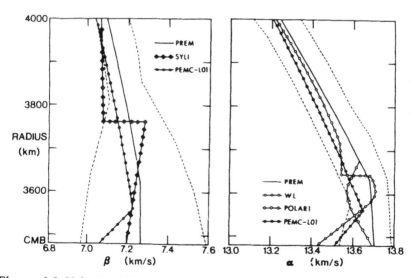

Figure 6.2 Velocity of *P* waves (α) and *S* waves (β) near the core-mantle boundary (CMB) for various Earth models (after Young and Lay, 1987).

the velocity of both *P* and *S* waves decreases as the CMB is approached, others that the velocities remain essentially constant, and others that the velocities continue to increase slightly (Figure 6.2). Decreased velocity gradients have been used to infer a thermal boundary layer with a temperature drop of ~800°C. Long wave length variations of the seismic properties of the D″ layer have been inferred from tomographic inversion of P and S wave travel times and wave forms (section 2.6). Dziewonski (1984) analysed more than 500 000 travel-time residuals to obtain details of the P velocity structure near the base of the mantle and found that the low order spherical harmonic components show velocity variations of 1–1.5%. Other global P wave travel-time inversions have been carried out which, although agreeing with the long wave length characteristics found by Dziewonski, show poor agreement for shorter wave lengths. It must not be forgotten, however, that sampling of the deep mantle is not uniform so that some of the discrepancies between models are artefacts of the spherical harmonic expansions. Truncation of the expansions also provides strong smoothing of possible smaller scale velocity heterogeneities. Large scale variations have also been found in shear wave velocities in the D'' layer. In particular there is evidence for a ring of high shear-wave velocities with perturbations of several per cent beneath the circum-Pacific.

Long wave length topography on the CMB has been inferred by Morelli and Dziewonski (1987) from an analysis of PKP and PcP travel times. They found up to 10 km of relief with scale lengths of 3000–6000 km. The depressed regions tended to correlate with zones of higher seismic velocities in D'', suggesting that the topography is dynamically supported by the downwelling of cool mantle material. Doornbos and Hilton (1989) have examined the structure of the CMB using data from seismic waves reflected from the bottom side of the core (PKKP), in addition to waves refracted through the CMB (PKP) and waves reflected from the CMB (PcP). All of the earlier analyses used PKP and PcP data only, and Doornbos and Hilton point out that, although all these earlier studies show considerable lateral heterogeneity and topography, the results do not agree amongst themselves. Using the additional PKKP data, Doornbos and Hilton found that the CMB is relatively smooth compared with previous results, the strong lateral variations inferred from the PKP and PcP data alone being attributed in part to structure above the CMB. Their preferred model involves both topography and a laterally varying boundary layer, although the average layer thickness is not constrained by the present data.

Several investigators have used wave form modelling to look at special localized regions of the lower mantle. Differences have been found in the shear velocity structure beneath the Pacific as compared to that under surrounding areas. It is unlikely that this can be accounted for by a thermal boundary layer, suggesting that localized compositional stratification or a phase change is responsible. From an analysis of short-period diffracted phases from localized regions, Young and Lay (1989) have also found strong lateral gradients in the average P velocity structure in D″.

6.3 EVIDENCE FROM GEOMAGNETISM

As is the case with the interpretation of travel-time data, inversions of geomagnetic surface data to estimate the flow in the OC are controversial and inconsistent. Bloxham and Gubbins (1985, 1987) have mapped the Earth's magnetic field on to the CMB and found that some features have remained essentially static for several centuries (e.g., in the Pacific hemisphere), whilst others have shown fairly rapid changes (e.g., in the Atlantic hemisphere) (section 5.4). They also found a correlation between these different regions and variations in deep mantle seismic velocities. They therefore suggested that the mantle controls flow in the OC by thermal coupling. Regions of upwelling material in the core are associated with seismically slow (presumably hot) regions in D″, and core downwellings are associated with seismically fast (cold) regions in the mantle. Gubbins and Richards (1986) have also looked at the mechanical and thermal effects that topography on the CMB would have on flow in the core and on the geodynamo. If there is topographic relief on the CMB, the isotherms will no longer coincide with gravitational potentials so that there will be both a lateral temperature gradient as well as mechanical interaction with the flow. It has been suggested that reversals of the Earth's magnetic field may be the result of slowly changing temperature variations in the mantle, modulating the core flow regime. This will be discussed later in section 6.7.

6.4 EVIDENCE FROM MINERAL PHYSICS

Experiments with the laser-heated diamond anvil cell have provided new information on the material properties of the lower mantle mineral phase, silicate perovskite, as well as on melting relationships in the Fe–O, Fe–S systems relevant to the OC. The melting temperature of silicate perovskite places important bounds on the mantle geotherm. Heinz and Jeanloz (1987) found that the melting point increases from 3000 K to more than 3800 K over the pressure range 30–130 GPa. Estimates of mean lower mantle adiabats range from 2600–3100 K (Jeanloz and Morris, 1986) so that any major temperature increase in D″ could approach the solidus with the possibility of partial melting.

These experiments have also indicated that chemical reactions between the mantle and core may occur. Knittle and Jeanloz (1986, 1989) investigated the reaction of liquid iron with the high pressure perovskite phase of (Mg, Fe) SiO_3. They found that chemical reactions occurred at pressures >70 GPa and temperatures above 3700 K. X-ray diffraction studies of quenched samples showed that the reaction products included SiO_2 stishovite and iron alloys (Fe, Mg)$_x$ O and Fe_xSi_y. These experiments suggest that similar chemical reactions should occur at the CMB over geological time creating chemical heterogeneities of silicate-rich and iron-alloy rich zones in the D$''$ layer. The iron enriched zones would be accompanied by a large increase in the electrical conductivity. Knittle and Jeanloz further suggested that any heterogeneities at the top of the core would be rapidly dispersed by the flow of liquid in the OC.

As discussed in section 3.4 there is no concensus on what is the light alloying element in the OC although S and O are still the most favoured candidates. However they have markedly different effects on the alloy melting temperature. At high pressures, O increases the melting temperature of an Fe–FeO system relative to pure Fe, whilst S in an Fe–FeS system depresses the melting point from that of pure Fe. Although the actual eutectic composition of the OC is not known, these results indicate that the core temperature at the CMB is at least 3800 K. Some estimates put it as high as 4400 K. These high temperatures demand a temperature drop of at least 700 K across a thermal boundary layer in D$''$. There is of course no reason why O and S could not both be present in the OC. Preliminary results indicate that even small amounts of S in the Fe–FeO system would drastically depress the melting temperature.

Further experiments at high pressures and temperatures by Knittle and Jeanloz (1991b) reinforce their earlier suggestion that chemical reactions at the CMB lead to chemical heterogeneities in the D$''$ layer and the incorporation of light alloying components (principally O but some Mg and Si) into the Earth's OC.

The thermal conductivity K of the lower mantle is usually assumed, in the absence of any firm data, to be approximately constant with a value of 4 W m^{-1} K^{-1}. This value supports the view that D$''$ is a thermal boundary layer. Brown (1986) has re-estimated the value of K in the lower mantle showing the relative effects of different geotherms, differing experimental constraints on the temperature and pressure dependence of K and of different values of other parameters in theoretical studies. Most of his models lead to substantially higher values of K in the lower mantle which, if substantiated, would imply that D$''$ is not a simple thermal boundary layer and that some changes in chemical composition from the rest of the lower mantle are necessary as well.

Davies and Gurnis (1986) have carried out numerical experiments to see what effect mantle convection would have on a thin layer of denser material at the base of the mantle. Their results showed that substantial lateral variations in the thickness of the layer developed. A model with a density contrast of ~2% and an initial uniform thickness of the denser layer of 100 km produced a discontinuous distribution with a maximum thickness of 230 km and bottom topography of several km in reasonable agreement with seismic observations. This would affect the heat flux out of the core with large lateral variations in the thermal boundary conditions which may in turn be relevant to long term changes in the geomagnetic field. The situation is further complicated, since it is possible that this boundary condition could change in a relatively short time (10^7–10^8 y) in response to a new influx of cooler mantle material.

The D″ layer may well have a more complex structure – a thin (<100 km) basal, low-velocity, high attenuation zone that can be explained as a thermal boundary layer (with a temperature drop of 800 K) and, superimposed on this, another ~200 km layer of sub-ducted slab material defined by increased S wave velocity (Lay and Helmberger, 1983). Since the viscous relaxation time for un-supported topography at the CMB is <10^6 y, a dynamic mech-anism is needed to sustain it and account for its lateral scale and regional variability. Loper and Eltayeb (1986) have studied the stability of a D″ thermal boundary layer and shown that, if the viscosity drop across the layer is >10^3, convection can take place confined to the low viscosity sub-layer. Olson *et al.* (1987) have carried out numerical experiments which confirm this and shown that small scale convection cells develop within the low-viscosity sub-layer on a time scale of tens of Ma and with horizontal scales of tens of kilometres. Groups of these convection cells merge to form plumes also on a timescale of tens of Ma. If the D″ layer contains significant compositional heterogeneity, it would inhibit plume escape and increase this time. Boundary layer convection can also support small scale roughness on the CMB with half widths as small as 20 km and amplitudes of ~2 km. The formation of plumes at the CMB and their tectonic significance will be discussed in section 6.5.

What can be concluded about the D″ layer? The mineral physics experiments have provided strong evidence for a thermal boundary layer at the CMB. However temperature estimates are getting un-comfortably high perhaps causing the layer to become intrinsically unstable. If so one would not expect to see a strong D″ region in the absence of any chemical layering. Evidence for chemical heterogeneity is provided by the seismic velocity discontinuities (in some models), by the lateral velocity gradients and by the experiments indicating chemical reactions between molten Fe and silicate perovskite. There is no reason, of course, why these two models should be mutually exclusive. Perhaps the best solution is that D″ is a heterogeneous chemical boundary layer embedded in a thermal boundary layer. Thermal instabilities in the form of plumes as well as large scale mantle convection may reorganize and entrain the compositional heterogeneity. Both flow regimes could affect the topography on the CMB with a wide spectrum of scale lengths.

Processes that could potentially contribute to chemical boundary layer formation at the CMB include the accumulation of 'dregs' at the base of the mantle (e.g., subducted oceanic crust), the accumulation of 'slag' at the top of the OC (e.g., from IC differentiation), chemical reactions at the interface, and primary chemical layering as a result of inhomogeneous accretion and/or core formation. This system of chemical boundary layers may interact strongly with mantle convection and play a crucial role in coupling it to convection in the core.

6.5 PLUMES

Wilson (1963, 1965) introduced the idea of stationary mantle 'hot spots' across which lithospheric plates drift in order to explain the regular progression in age along the chain of volcanic Hawaiian islands. Later Morgan (1971, 1972) suggested that such hot spots were the surface expression of narrow plumes originating deep in the mantle. Stacey (1975) and Jones (1977) placed the source in the D″ layer at the base of the mantle and Stacey and Loper (1983) interpreted the D″ layer as a thermal boundary layer with a temperature increment across it of ~840 K. In a companion paper, Loper and Stacey (1983) showed that narrow, long lived plumes are a necessary consequence of lower boundary heating of a medium with strongly temperature dependent viscosity. The upward flow is confined to chimneys that are ~20 km in diameter at the base of the mantle and decrease in width with progressive upward softening and partial melting of plume material. The speed of flow up the axis of the plume is ~1.6 m y^{-1} at the base and 4.8 m y^{-1} at a depth of 670 km. The narrow rapid flow is a consequence of the low viscosity resulting from the high temperature of the flowing material – driven by the buoyancy of the hot material within the chimney. The plume flow is self-regulating with a slow radial *inflow* of mantle material constricting the chimney. Thus a plume cannot weaken and dissipate in the lower mantle as it is dynamically and thermally robust. Plumes thus transport hot material upward, presumably until some obstacle is reached. Loper (1985) suggests that the most likely obstacle is the cold rigid lithosphere. Some of the plume material can then occasionally break through, as at Hawaii, if the underlying plume is particularly vigorous. More likely, however, the plume material is deflected horizontally to form, and add to, the lithosphere. The partially molten material in the asthenosphere would cause seismic waves to have lower velocities, providing an explanation of the seismic low velocity zone.

The effects of temperature dependent viscosity and heating in the core-mantle transition zone would tend to lead to boundary layer instability and the formation of thermal plumes. A sufficiently large viscosity contrast across D″ ($\sim 10^4$) with resulting plume formation could result in significant short wave length (20–50 km) dynamic topography on the CMB. If there is chemical heterogeneity in D″, it would substantially modify the dynamical behaviour of the region, including plume instability and dynamic topography. Many investigations have been carried out on the effect of concentrations of chemical heterogeneity near the base of the mantle. Most of the models suggest that if heavy chemical dregs collect in D″, they will be swept into laterally discontinuous aggregations by deep mantle convection, with a tendency for them to be concentrated beneath mantle upwellings.

Griffiths and Campbell (1990) have carried out laboratory experiments using clear glucose syrup to study stirring and structure in mantle plumes. They showed that a large amount of surrounding fluid can be stirred into the plume as it rises. Theoretical calculations indicate that similar entrainment will occur in mantle plumes if they rise sufficiently far from their source region. Any compositional differences, however, are only stirred, not mixed, due to the negligible rate of chemical diffusion. On the other hand, the effects of stirring in the mantle will determine the size, velocity, temperature, and composition of the heads of starting plumes. If a plume originates at the CMB, Griffiths and Campbell estimate that its head is likely to have a diameter ~1000 km when it reaches the upper mantle. It will collapse rapidly beneath the lithosphere, producing a large thermal anomaly 2000 km or more across. In a later paper, Campbell and Griffiths (1990) suggest that plumes originating at the CMB can account for the lateral extent of volcanism and uplift associated with continental flood basalts.

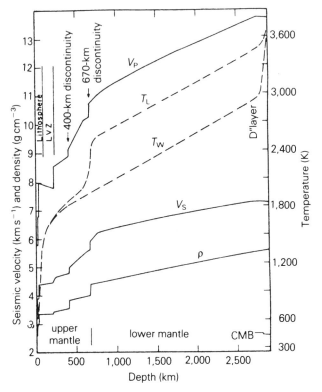

Figure 6.3 Profiles of the spherically averaged seismic wave velocities V_P and V_S, and density ρ through the mantle for Earth model PREM (solid lines). Possible mantle geotherms (dashed lines) are also shown, including the lithospheric and D″ layer thermal boundary layers. Geotherm T_W is appropriate for whole-mantle convection, T_1 for layered convection (after Olson *et al.* 1990).

6.6 PLATE TECTONICS AND MANTLE CONVECTION

In order to discuss possible interactions between the core and the mantle, it is necessary first to mention another, but far less dramatic division in the Earth – a transition region dividing the mantle into an upper and a lower part. Seismic data show two steps in the velocity depth curves at depths of about 400 and 670 km with a much smaller step at about 1050 km (Figure 6.3). These steps are accompanied by corresponding jumps in density. It is now generally believed that this transition region represents phase changes which involve changes in structure and density, but not composition. At the pressures and temperatures corresponding to a depth of ~400 km, laboratory experiments have shown that olivine (the principal mineral in peridotite) transforms to a more closely packed spinel structure (Akogi *et al.*, 1989). Laboratory experiments have shown a further breakdown to denser structures at a depth of 670 km where γ-olivine (spinel) transforms to perovskite and magnesiowüstite (Ito and Takahashi, 1989). It should be pointed out, however, that some researchers (e.g., Anderson and Bass, 1986) argue for chemical changes at these discontinuities.

No attempt will be made to discuss in any detail plate tectonics which has revolutionized geological thinking over the last few decades. In spite of the enormous literature on the subject, there are still some questions that remain unanswered, or at least are controversial. Are the plates active agents driving convection or are they passive layers riding on mantle convection cells? What is the structure of convection within the mantle – is it two-stage or mantle wide? This latter question is tied up with the nature of the 670 km discontinuity – does it represent a phase change or a chemical boundary? Two main approaches have been made to the problem of what is the driving force for plate motions. One concentrates on plate tectonic studies which generally conclude that the primary driving force for mantle convection is the negative buoyancy force of cold subducted slabs. The other approach is to consider thermal convection, the mantle being heated both internally by radioactive decay and from below by heat from the core. In this view the plates appear passively as a cold thermal boundary layer resting on an actively convecting mantle. I prefer the model of whole-mantle convection as developed by Loper (1985) for example which combines some features of both approaches. It is similar to the model of Richter and McKenzie (1978) in which mantle convection is driven by the negative buoyancy force of the descending slabs, except that in Loper's model, viscosity plays a much more important role. The structure of the convection is governed principally by viscosity not density. Differences in density drive convective motions, but the pattern of the motions is governed by differences in viscosity.

The viscosity of solid material is very temperature dependent e.g., the viscosity of mantle material at a temperature of 400°C is as much as a factor of 10^{10} greater than that at 1400°C. The rigidity of plates is just a consequence of being cold. For laboratory experiments, Lyle's Golden Syrup is often used. Its viscosity varies by a factor of 10^7 between −10°C and 60°C. This extreme temperature dependence of the viscosity leads to two types of mantle convection – rising and sinking of plumes and plate motions, with no obvious association between the two. The rising and sinking plumes form a slowly changing convective pattern with velocities less than 10 mm/y, whilst plates can move with velocities up to 200 mm/y. Plate motions generally have no deep structure associated with the upwelling part of the circulation. They produce magnetic lineations and large bathymetric and heat flow anomalies. The high heat-flow at ridges results from plate separation and not from a deep convective plume. The downgoing part of this flow extends to great depths in the form of sinking slabs in the subduction zones and produces large gravity anomalies. The other convective pattern is like that which occurs in a pan of water heated on a stove. Both rising and sinking limbs extend through the whole depth and both are associated with gravity anomalies, though heat flow anomalies are small.

The thermal structure of deep mantle plumes has been discussed by Loper and Stacey (1983). The convective pattern due to the heat flux from the core consists of a limited number of discrete narrow plumes rising from a thermal boundary layer (the D'' layer) at the base of the mantle. There is relatively little motion in the bulk of the lower mantle, i.e., the lower mantle does not overturn in response to heating from below. On the other hand, general mantle convection is driven by down-going, cold lithospheric slabs. The thermal structure of the mantle is determined by the interaction of these two convective flow patterns – the D'' plume flow pattern which cools the core and the slab-driven flow pattern which cools the mantle. This is illustrated in Figure 6.4. Below some level $r = r_p$, the mass flux F of the bulk of the mantle is negative, and the bulk of the mantle descends towards the D'' layer. Above $r = r_p$, $F > 0$ and there is general upward motion. In this model, there is no general overturning of the mantle, thus allowing the preservation of distinct chemical reservoirs in the upper mantle. One argument that has often been put forward for two layer convection in the mantle is the 1700 Ma mean age or 'residence time' of material in the upper mantle inferred from lead isotopes in oceanic crustal rocks. This would seem to indicate that homogeneous convective mixing of upper and lower mantle material does not occur. Loper's (1985) model answers this objection.

Silver *et al.* (1988) have suggested an alternative model to explain the conflicting evidence. They propose that there is a compositional boundary between the upper and lower mantle, but that the density difference attributable to the change in composition is <2%, the rest being due to phase changes. In this picture, the boundary acts as a selective filter, or 'semipermeable membrane', which permits the ~100 km oceanic lithospheric slab to penetrate into the lower mantle, but preventing the downward flow of the entire upper

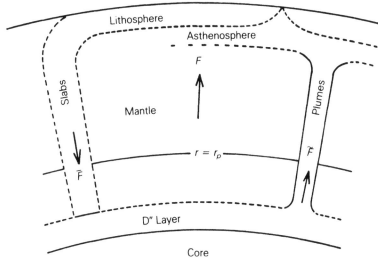

Figure 6.4 Different mass fluxes in the mantle. The slab mass flux $\bar{F}$, the plume mass flux $\tilde{F}$, and the mass flux F of the bulk of the mantle. $\tilde{F}$ is independent of radius, $\bar{F}$ is zero at the base of the mantle, and F is zero at some level $r = r_{\mathrm{p}}$ (after Loper, 1985).

mantle. The garnet-rich composition and low temperature of the slab will give it high enough density to sink into the lower mantle, whereas typical, hotter and less dense upper mantle material will continue to float above the denser, lower mantle (Figure 6.5). Eventually heating by surrounding hot lower mantle material will lower the density of the descending slab sufficiently to cause it to float up into the upper mantle again. Silver *et al.* estimate that the characteristic time for the round trip of the slab material through the lower mantle is quite short (~200 Ma). Thus the slab will be in the upper mantle most of the time and the ~1700 Ma upper mantle residence time as indicated by the lead isotope data will not be seriously affected. In their model, there is neither wide-scale mixing of geochemical reservoirs nor whole mantle overturn. It must be stressed once again that there is no general agreement on the pattern of convection in the mantle. The models of Loper and Silver *et al.* are but two that have been proposed.

It is worthwile considering in some detail the effect that a phase transition has on convection in the mantle. Two effects have to be considered – the coupling of latent heat with the ordinary thermal expansivity of the material and the effect of latent heat and advection of ambient temperature on the position of the phase boundary. For the olivine-spinel boundary at 400 km, the phase change is

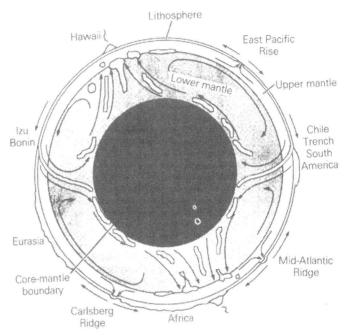

Figure 6.5 Schematic drawing of convection in the Earth as proposed by Silver, Carson and Olson. The model satisfies both the 1700 Ma residence time of material in the upper mantle and the evidence for lower mantle slab penetration. Because of a ~2% density increase from the upper to the lower mantle, slab material will descend into the lower mantle with little entrained flow and rapidly return to the upper mantle as a result of its intrinsic positive buoyancy while in the lower mantle. Once in the upper mantle, this material loses its compositional buoyancy and mixes in with the rest of the material (from Carnegie Institution of Washington, Year Book 87).

exothermic (i.e., the chemical reaction occurs with a liberation of heat) and the Clapeyron slope $\gamma = dp/dT > 0$. On the other hand at the 670 km discontinuity where the spinel structure of olivine $(Mg, Fe)_2SiO_4$ transforms into a mixture of perovskite $(Mg, Fe)SiO_3$ and magnesiowüstite $(Mg, Fe)O$, the reaction is endothermic, i.e., it occurs with an absorption of heat.

Consider first ordinary thermal expansivity and latent heat. A parcel of fluid moving downward through the olivine-spinel (exothermic) phase change is heated by the release of latent heat. It thus experiences an upward stabilizing buoyancy force, because of ordinary thermal expansion. In contrast, a parcel moving downwards through the spinel-postspinel (endothermic) phase transition cools by supplying latent heat, contracts due to thermal expansivity and is subject to a downward destabilizing body force.

Consider next the effect of the release or disruption of latent heat on the displacement of the phase boundary. Because of the heat liberated when material moves downward through the olivine-spinel phase change, the local temperature tends to increase and since the phase change interface must remain on the Clapeyron curve, the boundary is displaced downwards to higher pressure $(dp/dT > 0)$, i.e., the phase change has a destabilizing effect. Conversely for material moving downwards through the olivine-post olivine phase change, heat is absorbed, the local temperature tends to decrease, and the boundary is displaced upwards to lower pressures $(dp/dT < 0)$, i.e., the phase change has a stabilizing effect. Similar results apply to material moving upwards through a phase transition. These results are summarized in Table 6.1.

Christensen and Yuen (1984) have considered the influence of a chemical or a combined chemical and phase boundary in numerical (2-D) convection models which included subducting slabs. They found that the slab would be stopped at the boundary and bent sideways when the compositional density difference exceeds ~4–5%. The impact of the slab causes a depression of the boundary between upper and lower mantle of 100–200 km. It is not, however, very likely that the compositional density difference can exceed a few per cent. When it is <4%, the slab was found to

Table 6.1.

	Phase Boundary Distortion		Ordinary Thermal Expansivity + Latent Heat
	Advection of Ambient Temperature	Latent Heat	
Olivine-Spinel (Exothermic)	Destabilizing	Stabilizing	Stabilizing
Spinel-Post Spinel (Endothermic)	Stabilizing	Stabilizing	Destabilizing

Summary of the stabilizing and destabilizing effects of exothermic and endothermic univariant phase changes associated with phase boundary distortion and ordinary thermal expansivity plus latent heat (after Schubert *et al.*, 1975).

plunge deeply into the lower mantle and with <2%, it would reach the CMB and lead to large scale mixing of upper and lower mantle material. Kincaid and Olson (1987) found semi-quantitatively the same behaviour in laboratory experiments using corn syrup to model slabs in a stratified convecting system.

Christensen (1989) has carried out further numerical calculations and shown that, if the 670 km discontinuity is caused by an isochemical phase transition, it has to have a Clapeyron slope $dp/dT \leq -67$ MPaK^{-1} to prevent convection currents from crossing. This value is improbably low. If the discontinuity represents a chemical boundary, the intrinsic density difference has to exceed 3% to prevent subducted lithospheric slabs from penetrating deeply into the lower mantle – this condition is also possibly hard to meet. Christensen thus prefers whole mantle convection.

Ito *et al.* (1990) have developed a new calometric technique to obtain thermochemical data for very small amounts of $MgSiO_3$ perovskite obtained by synthesis at 25 GPa in a multi-anvil apparatus. They found that the thermodynamics of the reaction Mg_2SiO_4 (spinel) to $MgSiO_3$ (perovskite) and MgO (periclase) clearly has a negative pressure temperature slope of -4 ± 2 MPa K^{-1}. This is similar, but somewhat smaller in magnitude than the value $(-6$ MPa K$^{-1})$ obtained by Christensen and Yuen (1984) to completely inhibit convection. Ito *et al.* thus concluded that whole mantle convection cannot be ruled out by the thermochemical properties of $MgSiO_3$ perovskite. They warn, however, that their conclusion depends critically on uncertainties in the numerical analysis of convection and on the effects of other constituents on the phase boundary slopes.

Different conclusions have been reached by Jeanloz and Knittle (1989). They found that for plausible bounds on the composition of the upper mantle (ratio of Mg to Fe and Mg components >0.88) and temperature in the lower mantle (>2000 K) the high pressure mineral assemblage of upper mantle composition is at least 2.6 (± 1)% less dense than the lower mantle over the depth range 1000–2000 km. Thus a model of uniform mantle composition is incompatible with existing mineralogical and geophysical data. They believe that the mantle is stratified with the upper and lower mantle convecting separately. In this case of layered mantle convection, the temperature of the lower mantle would be at least 500 K higher than in the case of single layer convection, because of the presence of an extra thermal boundary layer near a depth of 700 km. The result is to increase the required contrast in intrinsic density between the upper and lower mantle to about 5 (± 2)%. Jeanloz and Knittle admit that the layering could be 'leaky' and that there could be a finite (but incomplete) amount of intermixing. Both intermixing and an increase in the viscosity of the lower mantle would tend to obscure the presence of a thermal boundary layer between the upper and lower mantle.

Bercovici *et al.* (1989a,b) have carried out 3-dimensional spherical calculations of convection in the Earth's mantle and come to somewhat different conclusions from Loper (1985). Two-dimensional models of mantle circulation, although useful in exploring the vertical structure and heat flow characteristics of convection, cannot help model the horizontal geometry of trenches, ridges and hot spots. Bercovici *et al.* assume whole mantle convection, but take into account internal heating due to the decay of radionuclides (^{238}U, ^{235}U, ^{232}Th and ^{40}K) and secular cooling. Thermal history models of the Earth indicate that 70–80% of the heat flow is internally generated. However, as discussed in section 4.3, the high-pressure melting temperatures of Fe imply that temperatures at the CMB are much higher than earlier estimates with a much larger contribution of heat coming from the core. Bercovici *et al.* therefore considered three cases: heating entirely from below with isothermal boundaries; heating entirely internal, the lower boundary being insulated and the upper boundary isothermal; and finally models with both internal heating and heating from below, both boundaries being isothermal.

The calculations were carried out at a Rayleigh number Ra $\sim 100\,\text{Ra}_{cr}{}^*$. This is about one order of magnitude lower than the range of Rayleigh numbers for the Earth's whole mantle. Bercovici *et al.* do not expect, however, that the qualitative nature of the convective patterns (i.e., the structure and geometry of the upwelling and downwelling regions) will change significantly. For the case where heating is entirely from below (i.e., from the core), the convective pattern is dominated by large cylindrical upwelling plumes surrounded by a network of downwelling sheets, with thicknesses of the order of 1000 km (see Figures 6.6a and 6.7a). The maximum velocity in the upwelling regions was always greater than in the regions of downwelling. In the second case in which the mantle is heated entirely from within, the horizontal convective pattern has both cylindrical and long sheet-like downwellings and broad regions of weak upwelling (Figure 6.7b). The cylindrical downwellings were transient, whereas the downwelling sheets had great horizontal extent and were long-lived. Moreover in contrast to the first case, the largest velocities occurred in the downwelling

---

*Ra is a non-dimensional measure of the vigour of convection – see section 4.6.

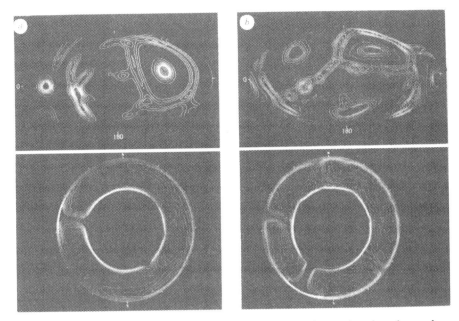

Figure 6.6 Numerical calculations of three-dimensional thermal convection in a spherical shell. The top panels show contours of vertical velocity at mid-depth. The bottom panels show entropy (potential temperature) contours in a cross-section of the spherical shell along a plane of constant longitude. Case (a) shell entirely heated from below. Case (b) shell half heated from below and half internally heated. The original diagrams were colour-coded and do not reproduce well in black and white. Interpretation is given in the text (after Bercovici *et al.* 1989b).

rather than in the upwelling regions. For the third case in which the mantle is both internally heated and heated from below, the convective pattern was similar to that of the first case (Figure 6.6b). As the proportion of internal heating is increased, the number of upwelling plumes increases and downwelling sheets become more vigorous. With any amount of heating from below, the entire convective pattern during its evolution is anchored to the upwelling plumes.

The most striking feature of the three models is that, although the patterns and nature of convection vary with changes in the mode of heating, the dominant style of downwellings is in the form of long, linear sheets. Descending slabs have usually been thought to occur because the lithospheric plates that become unstable and sink are rigid. The work of Bercovici *et al.* has shown that the downwelling sheets or slabs are the natural consequence of the

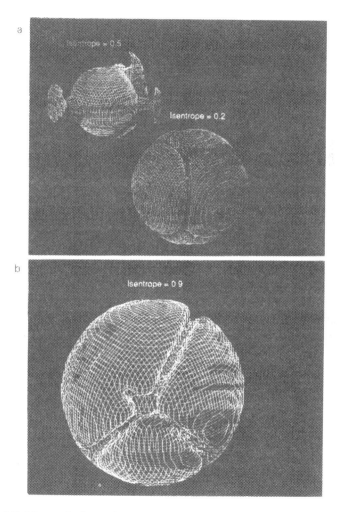

Figure 6.7 Numerical calculations of three-dimensional thermal convection in a spherical shell, showing three-dimensional surfaces of constant entropy. The values of the isentropes shown correspond to a non-dimensional measures of the entropy. Case(a) Shell entirely heated from below. The protrusions in the hotter isentrope ($s = 0.5$) are cylindrical upwellings. The canyons in the colder isentrope ($s = 0.2$) show the sheet like downwellings. Case(b) Shell entirely internally heated. In this case, because upwelling is present as a weak background flow, even the hot isentrope shown ($s = 0.9$) displays only the long linear canyons that delineate the downwelling sheets (after *Bercovici et al.*, 1989b).

dynamics of three-dimensional convective flow in a spherical geometry. This is in contrast to Loper's model. Moreoever cylindrical mantle plumes are the only form of upwelling and are not just secondary convective currents separate from large scale mantle circulation.

One further interesting result of their studies is that upwelling sheets that might be associated with mid-ocean ridges did not occur in any of the models with some heating from below. Moreover hot spots did not in general coincide with mid-ocean ridges. Thus mid-ocean ridges are not a consequence of upwelling from the deep mantle – rather the ridges are passive phenomena resulting from the tearing of surface plates by the pull of descending slabs as had been suggested much earlier by Lachenbruch (1976). Seismic tomography has also shown that slow velocities and therefore presumably hot regions beneath mid-ocean ridges do not extend deeper than ~350 km (Woodhouse and Dziewonski, 1984). Bercovici *et al.* conclude that the relatively large number of plumes in the Earth's mantle (inferred from the number of hot spots) and the evidence that downgoing slabs are vigorous enough to penetrate the 670 km discontinuity into the lower mantle suggest that the mantle is predominantly heated from within.

These calculations of Bercovici *et al.* (1989a,b) of three-dimensional convection in a spherical shell have been extended by Glatzmaier *et al.* (1990) to higher Rayleigh numbers Ra and at a resolution a factor of two higher in both latitude and longitude. Their value of Ra is ~200 times greater than Ra_{cr} for the onset of convection, but is probably still ~10 times smaller than the average value for the Earth's mantle. Their results (with 50% and 80% internal heating) are more temporally complex and definitely chaotic (section 1.4). In the earlier calculations of Bercovici *et al.* convection patterns tended to be anchored to the upwelling plumes which were relatively steady in number and location. In the later calculations of Glatzmaier *et al.* the upwelling plumes are more variable owing to their interaction with chaotic downwelling. However the sheet-like downwellings seen in the previous calculations continue to dominate the structure of convection. This suggests that the descending slabs in the Earth are the main-driving force for plate tectonics and mantle convection. The chaotic, subduction like downfolds may explain the apparently random dispersal and aggregation of the continents.

The pattern of thermal convection in a spherical shell is geometrically similar to that imaged by global seismic tomography. The pattern is also consistent with tomographic models predicting the distortion of the surface gravitational field caused by lower

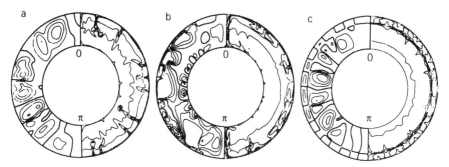

Figure 6.8 Isocontours of streamfunctions and temperature fields for three values of the Clapeyron slope γ. No influence of the phase change is evident for γ = 0 (case a) whereas a layered structure appears for γ = −4 MPaK^{-1} (case c). For γ = −2 MPaK^{-1} (case b) there is a mixture of whole mantle and layered convection (after Machetel and Weber, 1991).

mantle flow. It has been argued that mantle plumes originate in the core-mantle transition region, primarily as a result of boundary layer instabilities, but small, plume-like features in the lower mantle have not yet been resolved by seismic tomography. Plumes may originate from upper mantle transition zone boundary layers, or even from temperature-dependent phase boundaries.

Further light on the possible effect on mantle convection of a phase change at a depth of 670 km has been given in numerical studies by Matchetel and Weber (1991). Their model includes a phase change at 670 km as well as a spherical, axisymmetric geometry, compressibility and internal heating. Figure 6.8 shows their results for three values of the Clapeyron slope γ. For γ = 0 convection is not affected by the phase change (Figure 6.8a). For γ = −4 MPa K^{-1} penetration of sinking currents into the lower mantle is opposed leading to layered convection between the upper and lower mantle (Figure 6.8c). Finally for γ = −2 MPa K^{-1} (a value which is in the range of experimental data and theoretical estimates) local intermittent mixing between the upper and lower mantle is observed (Figure 6.8b). Convection is layered almost everywhere except in a few places where upper cells can penetrate into the lower mantle. For this case Machetel and Weber estimate the time average transfer of mass from the lower to the upper mantle over 1000 Ma to be ~10% while layering exists, rising to ~80% for short periods during mixing which occurs approximately every 500 Ma. The resulting time averaged mass transfer (over 1000 Ma) is 22%. Their model goes some way to reconciling the conflicting geophysical evidence for both whole mantle and layered convection.

An excellent review article on the structure of convection in the mantle has been given by Olson *et al.* (1990) and on future goals and trends in studies of the Earth's deep interior by Lay *et al.* (1990).

6.7 MASS EXTINCTIONS, REVERSALS OF THE EARTH'S MAGNETIC FIELD AND PLUMES

Two other problems – reversals of the Earth's magnetic field and mass extinctions, have been suggested by some authors to be connected with events originating at the CMB. Jones (1977) was the first to suggest that long term variations in reversal frequency may be the result of fluctuating temperatures at the CMB caused by intermittent breakdown of a static D'' layer. His model assumes that the geodynamo is driven by thermal convection in the OC and that the D'' layer is a thermal boundary layer. McFadden and Merrill (1984) disagree with the details of Jones' model, although conceding that his general ideas may be correct. In Jones' model, the thermal boundary layer becomes unstable and breaks down by the formation of blobs or plumes. The destruction of such a boundary layer takes up only a small fraction of the time in each thermal cycle, the buildup of a super adiabatic temperature gradient by thermal conduction requiring the longer time. Thus there would be a relatively rapid decrease in mean polarity length following the long Cretaceous normal interval with a subsequently longer period of time in which the mean polarity lengths *increase*. This is not what is observed. A *decrease* in the length of polarity intervals since the Cretaceous appears to have continued up to $\sim 10\,\mathrm{Ma}$ ago (section 5.5).

Loper and McCartney (1986) have developed these ideas further. They first point out that the D'' layer is not static and, as was shown by Yuen and Peltier (1980) must actively convect. They assume that the rate of reversals is related to the rate of supply of energy to the dynamo, which, in their model, is directly related to the rate of cooling of the core which in turn is controlled by the D'' layer. When the D'' layer is thick, the temperature gradient across it is small, the energy supply to the dynamo is low so that it is in a quiet state with few reversals. On the other hand when the D'' layer is thin, the temperature gradient across it is large, the energy supply to the dynamo is greater so that it is in a more disturbed state with frequent reversals. It is interesting that Sheridan (1983), who has developed a model based on similar ideas, has come up with a correlation just the opposite to that of Loper and McCartney. In his model, reversal frequency is low during periods of plume eruption and high when plumes are absent.

The other role that has been suggested that events at the CMB may play is in being indirectly the cause of mass extinctions. Few topics have created as much debate in recent times as the cause of mass extinctions at the Cretaceous/Tertiary (K/T) boundary. The literature on the subject is enormous and no attempt will be made to try and review it. A conference on Global Catastrophes in Earth History was held at Snowbird, Utah in 1988. It was clear that there were not only disagreements in the interpretation of the K/T boundary data, but that there were also fundamental differences in observations of the chemical and physical data from some of the sites. A report on the conference has been given by Chapman (1989). Albritton (1989) has given a useful short summary of the K/T debate. The basic question is whether the cause was extraterrestrial resulting from the impact of an asteroid or a comet, or whether it was due to volcanism. The impact hypothesis caught the attention of the scientific world by the publication of a paper by Alvarez *et al.* (1980) which reported anomalously high levels of Ir at the K/T boundary in deep-sea limestones exposed near Gubbio, Italy and in Denmark. Since then the Ir 'spike' has been found over a large portion of the globe, in both marine and terrestrial environments. Additional evidence in support of the impact model is the discovery of anomalous concentrations of other trace elements and isotopes and quartz grains showing shock metamorphism. The case for an internal cause has been strengthened by the dating of the Deccan flood basalts at ~65 Ma, the date of the K/T boundary, (Courtillot *et al.*, (1988)).

Crocket *et al.* (1988) found significant Ir enrichment over background in several clay layers surrounding the K/T boundary in Umbria, Italy. They concluded that a long period of intense volcanic activity explains the data better than the impact of an external body whose effect would be instantaneous. Rocchia *et al.* (1990) resampled the Gubbio K/T boundary studied by Alvarez *et al.* (1980) making both magnetostratigraphic and Ir measurements. They also found that the Ir concentrations stand out above the background level over almost 3 m of section which corresponds to ~0.5 Ma based on magnetostratigraphy. Although diffusion might account for part of the vertical extent of the Ir enriched interval, they are of the opinion that the most likely explanation is a protracted duration of the source of the Ir, on top of which the main (short-lived) K/T boundary anomaly stands.

Glass and Heezen (1967) pointed out that the great field of

tektites* covering Australia, Indonesia and a large part of the Indian Ocean fell about 700 000 y ago at about the time of the last magnetic reversal (the Brunhes-Matuyama chronozone reversal). They suggested that the fall of the body from which the tektites were formed killed the now extinct radiolaria and gave a jolt to the Earth, disturbing motions in the core and causing the dynamo to reverse. But as Bullard (1968) pointed out, it is 'difficult to believe that the fall of a large meteorite could selectively kill certain species of radiolaria all over the world and yet spare the kangeroos near its point of fall'. De Menocal *et al.* (1990) have examined the relative stratigraphic positions of interglacial stage 19.1 (determined by oxygen isotope studies), the Brunhes-Matuyama reversal and the widespread tektite layer in deep sea sediments. They found that the Brunhes-Matuyama reversal occurred 6 ± 2 kA after the stage 19.1 datum which in turn occurred 9 ± 3 kA after the tektite strewn field. (Figure 6.9). They concluded that there is no connection between geomagnetic field reversals, climate change and impact events.

Durrani and Khan (1971) and Glass and Zwart (1979) suggested later that there could be an association between the Ivory Coast microtektites and the onset of the Jaramillo normal polarity subchron about 0.97 Ma ago (section 5.5). This prompted Schneider and Kent (1990) to re-evaluate the palaeomagnetic stratigraphy of two critical deep sea cores containing Ivory Coast microtektites. They concluded that the event which produced them (the Bosumtwi impact crater in eastern Ghana? – see Shaw and Wasserburg, 1982) most likely occurred during the Jaramillo subchron, but ~30 000 y after its onset and ~40 000 y before its termination, thus giving no support for any causal relationship between the two events. Moreover the K/T boundary which as already discussed may mark a large impact event does not coincide with a polarity reversal (Lowrie and Alvarez, 1977). Furthermore there appears to be no record of tektites or microtektites associated with the times of other field reversals (there were more than 70 during the last 20 Ma).

The issue has been further complicated by the claim of some authors to have found a periodicity in mass extinctions of ~30 Ma (Raup and Sepkowski 1985, 1988; Rampino and Stothers, 1984), although such claims remain controversial and poorly supported statistically. Two hypotheses have been proposed to account for

---

* Tektites are small, glassy objects a few cm in size found in a few special areas of the world (e.g., Australia, the Ivory Coast, Czechoslovakia and North America. Microtektites (<1 mm in size) have also been found in deep-sea cores from the Indian and Pacific Oceans.

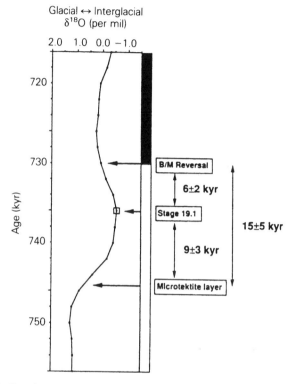

Figure 6.9 Graphic representation of the relative temporal offsets between the Brunhes-Matuyama reversal, stage 19.1 isotopic event and the Australian microtektite strewnfield layer (after de Menocal *et al.* 1990).

such a periodicity in mass extinctions. The first suggests that an undiscovered member of the solar system destabilizes the orbits of comets which then impact the Earth. The second involves the movement of our solar system above and below the galactic plane which again causes instabilities in the cometary cloud. There is no evidence to substantiate these hypotheses. Additional 'evidence' for a periodic bombardment of the Earth has come from crater ages, but such 'evidence' is very controversial. Grieve *et al.* (1985) concluded that, because of its inherent nature, there are considerable uncertainties in using time-series analyses of the cratering record, and that a variety of periods could be defined depending on which sample of the record was considered most representative. There is also no really convincing mechanism to explain how surface events could affect the reversal frequency of the dynamo. Even if an impact could trigger an individual reversal, it is unlikely to initiate a change in the frequency of reversals. Finally Raup (1985) has claimed a periodicity of ~30 Ma in the record of geomagnetic reversals which again has been disputed (e.g., Lutz, 1985). The net result of all these claims for periodicities is the suggestion that there is a correlation between large impacts on the Earth, mass extinctions and geomagnetic reversals. I find such a correlation extremely unlikely. Some of the arguments seem very tenuous, e.g., Muller and Morris (1986) suggest that dust from an impact crater and soot from fires would trigger a climatic change resulting in an ice age. The increase in the amount of ice would alter the moment of inertia of the Earth and hence its rate of rotation which in turn would disrupt the flow in the liquid OC and hence trigger a reversal.

For these reasons I will confine myself to causes internal to the Earth which have been proposed to explain some of the above phenomena, and, since we are interested in the deep Earth, I will return to the suggestion of Loper and McCartney (1986) that they may be governed by periodic instability of the thermal boundary D″ layer. Loper *et al.* (1986, 1988) carried out analogue experiments in the laboratory in which a layer of dyed water representing the heated material in the D″ layer is placed below viscous corn syrup representing the cold mantle. This initial configuration is dynamically unstable – a complete overturn is prevented by placing a silk membrane between the two fluids. Figures 6.10 and 6.11 illustrate some of the flows that were observed, which suggest that in the D″-plume system they may be unsteady and quasi-periodic.

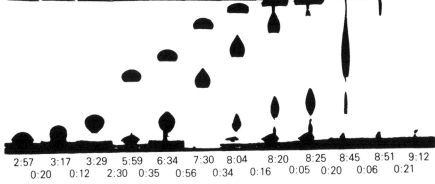

| 2:57 | | 3:17 | | 3:29 | | 5:59 | | 6:34 | | 7:30 | | 8:04 | | 8:20 | | 8:25 | | 8:45 | | 8:51 | | 9:12 |
| | 0:20 | | 0:12 | | 2:30 | | 0:35 | | 0:56 | | 0:34 | | 0:16 | | 0:05 | | 0:20 | | 0:06 | | 0:21 | |

Figure 6.10 This and Figure 6.11 each show a sequence of twelve photographs of one run of an experiment described in the text to study the evolution of mantle plumes. The elapsed time is shown immediately beneath each photograph in minutes and seconds, while the second line of numbers gives the time difference between adjacent photographs (after Loper and McCartney, 1986).

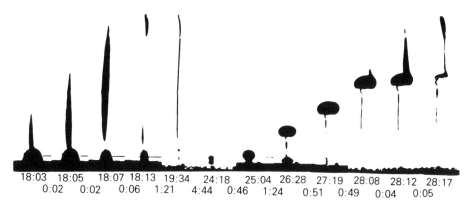

18:03	18:05	18:07	18:13	19:34	24:18	25:04	26:28	27:19	28:08	28:12	28:17
0:02	0:02	0:06	1:21	4:44	0:46	1:24	0:51	0:49	0:04	0:05	

Figure 6.11 During the time interval 12.16 to 17.00, a second plume was established and the first became inactive – this sequence is not shown. The first five photographs show a sharply peaked pulse of fluid which rapidly rises through the syrup. Between the fourth and sixth photographs, a steady plume prevailed. The last seven photographs show the growth of a third dome of fluid which rises and grows until it intersects an inactive plume conduit. This provides a pathway for the buoyant water to rise rapidly to the surface (after Loper and McCartney, 1986).

They estimated a thermal relaxation time of ~22 Ma which is reasonably close to the postulated 30 Ma periodicity in reversals. However the thermal conductivity of the D'' layer is not well known and their estimate of 22 Ma is very uncertain.

Any surface effects would depend on the amount of material and the history of its ascent. The laboratory experiments described above indicate that hot, low-viscosity material rises much faster within a pre-existing plume. This should cause an increase in non-explosive volcanism, because it can vent easily to the surface. If a diapir of low mantle material rises outside an established plume, it rises at a much slower rate, allowing it to grow much larger as it is fed by material from below resulting in a very explosive event. An example of such an event at the K/T boundary could be the Deccan flood basalts which at that time were situated over the Reunion hotspot. It must be emphasized that much of the plume model, like others, is highly speculative – we really know very little about what happens when a diapir of lower mantle material reaches the lithosphere under-surface. The existence of plumes rising from the CMB seems to be on a firm basis, but it is difficult to assess their effect on the surface of the Earth.

7

Epilogue – the cores of the other planets

7.1 INTRODUCTION

This book has been about the deep interior of the Earth and in particular its core. In Chapter 1 we discussed briefly the origin of the solar system insofar as it had a bearing on the origin of the Earth's core. It is thus natural to ask whether the other planets have cores. Since the main source of information on the Earth's interior comes from seismology and since only on the Moon have seismometers been installed outside the Earth, it is difficult to answer this question. Since we believe that the Earth's magnetic field is generated by dynamo action in the fluid OC, we can learn something about the interiors of the other planets from a knowledge of their magnetic field. This we have obtained in the last 25 years from measurements made by spacecraft. It has been found that the Moon and the terrestrial planets essentially have no magnetic field. In this regard it is interesting to speculate whether an IC is necessary for the operation of a dynamo and whether the smallness of magnetic fields presently observed on the Moon and other terrestrial planets is because these bodies do not now have a solid IC and a fluid OC. The growth of a metal rich IC causes the composition of the liquid OC to evolve towards the eutectic. Once the eutectic composition is reached, the gravitational dynamo will cease to function because the solid that freezes from a eutectic liquid has the same composition as the liquid.

Constraints on models of the interiors of the planets are also given by astronomical data such as their mass and mean diameter. The moment of inertia can also be estimated – quite accurately in the case of Mars and the Moon, but less so for Venus and Mercury. In addition some information has been obtained from theoretical studies on subsolidus convection in the terrestrial planets (see e.g., Schubert, 1979). Key issues are the power of solid-state mantle convection to remove substantial amounts of heat from the core, and the ability of light alloying materials to substantially lower the melting point of pure iron. Whether a differentiated model leads to

Table 7.1. Observations and interpretations of planetary magnetic fields
(after Stevenson, 1983)

	Observed surface field (Γ)	Interpretation
Mercury	$\sim 2 \times 10^{-3}$	Thin shell dynamo
Venus	$\leqslant 2 \times 10^{-5}$	Stable, fluid core; no dynamo
Earth	0.3	Dynamo maintained by innercore freezing
Moon	$\leqslant 2 \times 10^{-6}$	Stable fluid shell; no dynamo
Mars	$\leqslant 10^{-4}$	Stable, fluid (S-rich) core; no dynamo
Jupiter	4–10	Dynamo maintained by thermal convection
Saturn	0.21	Dynamo maintained by thermal convection

$\neq 1\ \Gamma = 10^{-4}T = 100\mu T$

a molten core persisting to the present depends critically on initial conditions, the values of the physical parameters assumed, and the possibility of solid state convection in the mantle.

It must be stressed at the outset that a determination of the constitution of the interiors of the Moon and planets is not unique. I believe that the models of Stevenson *et al.* (1983) are the best predictions that we can make at the moment. In their models they assume that core dynamos (if they exist) are driven by thermal and/or chemical convection, that radiogenic heat production is confined to the mantle, that mantle and core cool from initially hot states that are at the solidus and super-liquidus respectively and that any IC that may nucleate excludes the light alloying component which then mixes uniformly upwards through the OC. All their models assume whole mantle convection. Material parameters are chosen so that the present day estimates of heat flow, upper mantle temperature and IC radius are obtained for the Earth. They found that small changes in model parameters can result in completely fluid non-convecting cores (and hence no dynamo action), convecting fluid OCs with IC growth (with dynamo action), and almost solid cores with only thin OC fluid shells now remaining (and probably with no dynamo). A summary of their conclusions is given in Table 7.1 and in Figures 7.1 and 7.2. These will be briefly discussed later. In their models of the Earth, the IC is a late feature in its history. Since the Earth's magnetic field has existed for at least 3.5 Ma, Stevenson *et al.* (1983) suggested that the mode of powering the geodynamo may have changed during geologic time. Before IC formation, a thermal dynamo operated within the Earth with diminishing strength. After the onset of IC nucleation, gravitational energy would contribute to the power of the dynamo.

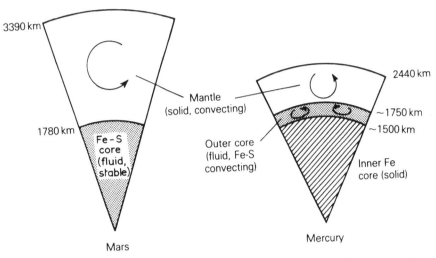

Figure 7.1 Schematic representation of probable present-day states for Mars and Mercury. Mars may have no IC if the core is sulphur-rich whereas Mercury's core may be mostly frozen because of very low sulphur abundance. The thin but vigorously convecting Mercurian fluid shell may be capable of magnetic field generation (after Stevenson, 1983).

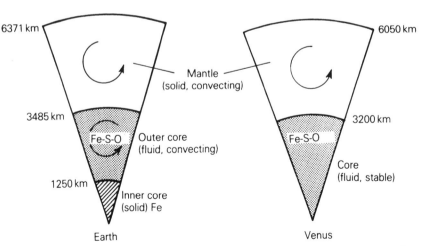

Figure 7.2 Schematic representation of probable present-day states for Earth and Venus. The lower pressures and higher temperatures in Venus may prevent IC growth. This would cause the core to be stably stratified and incapable of magnetic field generation (after Stevenson, 1983).

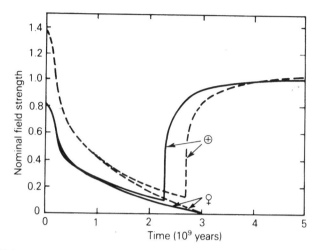

Figure 7.3 Nominal magnetic field strength through geological time for two models of the Earth and Venus (after Stevenson *et al.* 1983).

This should result in a steep increase in the nominal field strength. Since that time, the field strength should remain ~ constant or decrease slowly by about 10% in 2 Ga. (Figure 7.3). Hale (1987) has reported that palaeomagnetic results support such a scenario – palaeofield determinations indicate an abrupt increase in the magnitude of the magnetic field about 2.7–2.1 Ga ago.

7.2 MERCURY

Mercury is perhaps the most interesting of the inner planets in that it does appear to have a small magnetic field of internal origin. Information about its magnetic field was obtained from Mariner 10 which made three passes by the planet between March 1974 and March 1975. On the first pass a maximum field strength of 98 nT was observed at closest approach (723 km). The third pass was closer to the planet (327 km) and at higher latitudes and a magnetic field of over 400 NT was recorded. External current systems strongly affect the measurements made by Mariner 10 and there has been much controversy over the strength of its magnetic moment – it is at least three orders of magnitude less than that of the Earth. The field is too large to be caused by any external induction process and must be of internal origin. Stevenson *et al.* (1983) suggest that Mercury's core may be mostly frozen because of very low sulphur abundance and that the field is generated in a thin, fluid Fe–S O C (Figure 7.1). The thickness of the Fe–S shell varies between about

85 and 485 km depending on the amount of light constituent in the core. However the field strength should be larger than that observed if the dynamo operates similarly to that in the Earth. Stevenson (1983) later suggested that there simply may not be enough energy to power the dynamo to the level at which the Earth's dynamo works, or that temperature differences along the core-mantle interface could drive thermo-electric currents.

7.3 MARS

Mars has been visited by a number of spacecraft. Those that investigated its magnetic field were Mariner 4 which passed within about 13 200 km in July 1965 and the Soviet Mars 2 and 3 which passed within 1100 km in November and December 1971. Further magnetic observations of Mars were made in February 1974 by the Soviet Mars 5, whose nearest approach was 1800 km. There has been much controversy on the interpretation of the results. The general concensus is that Mars has no magnetic field – the upper limit of a Martian magnetic moment being only 2×10^{-5} of the Earth's magnetic moment. Dolgniov (1987) still believes however that Mars does have a very weak intrinsic magnetic field. (for a full discussion, see Russell, 1987). In 1988, the Soviet Union launched two more spacecraft to Mars; Phobos 1 and Phobos 2, named after Phobos, one of the two Martian moons. Phobos 1 was launched on July 7, but unfortunately contact was lost on September 1 before it reached its objective. Phobos 2 was launched on July 12 and successfully went into orbit around Mars. However after 57 days in orbit, contact with Phobos 2 ceased owing to an on-board computer malfunction. Phobos 2 made five elliptical orbits with a closest approach to Mars of 850 km followed by more than 100 circular orbits at an altitude of 6000 km. Phobos 2 carried two triaxial flux-gate magnetometers. Analysis of the data by Riedler *et al.* (1989) gives no conclusive evidence for an intrinsic magnetic field.

Since Mars rotates at a rate close to that of the Earth and is expected to have a core comparable in size to that of Mercury, it is reasonable to expect that it would have a magnetic moment intermediate between that of Mercury and that of the Earth, i.e., about three orders of magnitude greater than that observed. The very much weaker magnetic moment of Mars suggests that it has at most a very small fluid electrically conducting core. It is possible that Mars had at one time a significant magnetic field driven by thermal convection which gradually diminished in strength. Stevenson *et al.* (1983) found that Mars admits all possibilities for the state of its core. They prefer a Fe–S, stably stratified fluid core (Figure 7.1). This is in agreement with a consmochemically plausible sulphur content of 15% or more by weight and the absence of a solid IC provides an explanation for the lack of a substantial magnetic field.

Schubert and Spohn (1990) have extended the model of Stevenson *et al.* (1983) of the thermal history of Mars. They assume that Mars is initially hot and completely differentiated into a core and mantle. This assumption is based on convincing evidence that Mars is the parent body of a special class of meteorites, the SNC meteorites (e.g., McSween, 1984). Analyses of the elemental and isotopic composition of the SNC meteorites indicate that a Martian core formed within a few hundred Ma at most of the end of accretion (Chen and Wasserburg, 1986). The light alloying element of the core is assumed to be sulphur. It is the initial sulphur concentration x_s that is the principal parameter controlling the future evolution of the core and a Martian magnetic field. Schubert and Spohn show that IC solidification depends mainly on x_s and mantle viscosity. Relatively small, sulphur-poor cores would now be largely solid, whilst relatively large, sulphur–rich cores would still be largely liquid. Figure 7.4 shows the change in heat flux from the core for two values of x_s. The sudden change in the rate of core

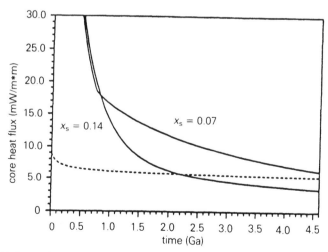

Figure 7.4 Evolution of the Martian core heat flux with time for core sulphur concentrations of 7 and 14% by mass. The dashed curve is the heat flux conducted along an adiabat in the Martian core. The sharp bend in the core heat flux curve for $x_s = 0.07$ that occurs at about 0.75 Ga marks the onset of inner core solidification (after Schubert and Spohn, 1990).

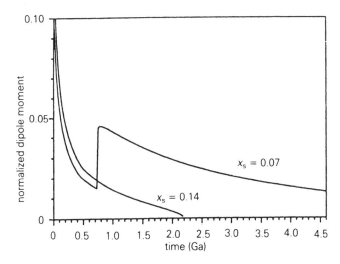

Figure 7.5 Normalized magnetic dipole moment as a function of time for two values of sulphur concentration in the Martian core. The sudden jump in the dipole moment that occurs at about 0.75 Ga for $x_s = 0.07$ coincides with the initiation of inner core freezing. No dynamo action is possible after about 2.2 Ga in the model with $x_S = 0.14$ since thermal convection in the core of this model does not occur subsequent to this time and there is no inner core freezeout in the model (after Schubert and Spohn, 1990).

cooling at about 750 Ma for $x_s = 0.07$ marks the onset of IC solidification. With $x_s = 0.14$, no solid IC is formed. The dashed curve in Figure 7.4 represents the heat flux conducted along the core adiabat. When the core heat flux falls below this curve, thermal convection can no longer be maintained in the core. With $x_s = 0.14$, this occurs at a time of ~2.2 Ga. After this time there would be no dynamo action and magnetic field generation. With $x_s = 0.07$, thermal convection is possible even beyond 4.6 Ga. In this case, convection in the core after 750 Ma is both compositional and thermal. This is further illustrated in Figure 7.5 which shows the Martian magnetic dipole moment normalized with respect to the Earth's present magnetic dipole moment as a function of time for the two values of x_s. Possible forms of mantle convection in both Mars and Venus are considered in the next section.

7.4 VENUS

Venus has also been visited by a number of spacecraft investigating the possibility of a planetary magnetic field. The first was Mariner 2 which passed within 41 000 km of the centre of the planet in

December 1962. Later in October 1967, Venera 4 approached to within 200 km above the surface of Venus and Mariner 5 about 10 150 km from the centre. Venera 6 impacted the planet in May 1969 and Mariner 10 passed within 11 900 km of the centre in February 1974. Venera 9 and 10 orbited Venus in October 1975, their closest approach being 1500 km. The results from all these missions, like those to Mars, have been somewhat controversial, although agreeing that Venus has at most a very small magnetic field. The last spacecraft to probe the magnetic field of Venus was the Pioneer Venus orbiter which went into orbit in December 1978, the nearest approach to the planet being 150 km. Dolginov still believes that Venus like Mars does have a very weak intrinsic magnetic field, but the general opinion is that an upper limit of a Venusian magnetic moment is only 10^{-5} of the Earth's magnetic moment (Russell, 1987 has again given a full discussion).

Venus, which is comparable in size to the Earth, rotates more slowly – about one quarter the rate of Mercury – and one would expect a moment at least two orders of magnitude greater than that observed. One can then ask whether this has always been the case. Stevenson *et al.* (1983) found that their models for the core of Venus admit a state similar to that of the Earth, but also a completely fluid, stable stratified core and a core that is mostly solid. They favour a completely fluid core (Figure 7.2). Solomatov and Zharkov (1990) have come to the same conclusion. Models with almost solid cores require implausibly low amounts of light alloying components. The absence of a significant IC in Venus is probably due to its slightly higher temperature and lower central pressure relative to the Earth. Perhaps Venus once had a substantial magnetic field that died ~1.5 Ga ago, and might eventually nucleate an IC, causing a renewal of the dynamo (Figure 7.3).

Schubert *et al.* (1990) have carried out calculations of three-dimensional mantle convection for both Venus and Mars similar to those of Bercovici *et al.* (1989) for the Earth. Their calculations are based on the assumption that plate tectonics is absent on both Venus and Mars, i.e., they are considered as one plate planets with rigid, immobile lithospheres. Thus mantle convection in Venus and Mars occurs beneath a rigid lid, and the boundary condition at the outer surface of the model spherical shell is the vanishing of the velocity. This is in contrast to the Earth where, because of mobile tectonic plates on the surface, the appropriate boundary condition is the vanishing of the shear stress. Their models for both Venus and Mars show that cylindrical plumes are the most prominent forms of upwelling if there is sufficient input to the mantle from the core. The form of the downwelling sheets is significantly affected by the rigid upper boundary conditions – the sheets are more irregular in their horizontal structure than when the top boundary is shear stress free. In models mainly heated-from-within, the downwelling sheets are also shorter and less temporally durable when the top boundary is rigid.

Schubert *et al.* (1990) attribute a number of the surface features of Venus to cylindrical upwelling plumes in the planet's mantle. Quasi-circular features (the equatorial highlands) with pronounced geoid anomalies are interpreted as surface swells over hot upwelling mantle plumes. Upwelling mantle plumes have also been suggested as the cause of coronae, oval shaped features, 150–600 km in diameter, uplifted several hundred metres above the surface and characterized by interior lava flows indicative of volcanism. Since mantle plumes arise from a thermal boundary layer at the lower boundary of the mantle, the existence of plumes in the mantle of Venus is indirect evidence that the planet has a core.

Major volcanic provinces on Mars, like Tharsis, may be the result of the cylindrical nature of upwelling mantle plumes, similar to hotspots on the Earth. There are no sheet-like upwelling features in the Martian mantle producing a pattern similar to the linear global systems of mid-ocean ridges on Earth. The rigid upper boundary condition on Mars inhibits the initiation of slab-driven tectonics. Any downwellings would be short lived and ineffective in disrupting the lithosphere. Thus the major tectonic features are hotspots and volcanic domes arising from mantle upwellings in the form of cylindrical plumes.

7.5 THE MOON

In 1959 the Soviet Luna 2 passed within 55 km of the Moon's surface and detected no magnetic field. This was confirmed by Luna 10 in (1966) and Explorer 35 in 1967. The results of the Apollo programme which followed were very surprising. Stable components of natural remanent magnetization of lunar origin were found in samples brought back by Apollo 11. Later lunar surface magnetometers at the Apollo 12, 14, 15 and 16 landing sites recorded surprisingly high local surface fields of tens and hundreds of nT. The samples brought back by the Apollo mission were collected from the lunar regolith so that the orientation in which they were formed is not known. Thus one cannot obtain information on the morphology of ancient lunar magnetic fields. However the intensity of the field can be measured. The results indicate that the field rose to a maximum of $\sim 100\,\mu T$ between about 3.9-3.7 Ga ago and then decayed to 5–10 μT until 3.1 Ga ago

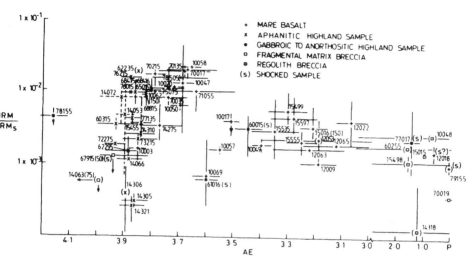

Figure 7.6 Plot of normalized NRM intensity against radiometric age for lunar basalts and breccias (after Cisowski et al., 1983). Horizontal scale is in 10^9y (aeons) and there is a discontinuity in the scale at 3.0 AE. The ordinate scale is proportional to the ancient field intensity. An NRM/IRM ratio of 2.0 × 10^{-2} corresponds approximately to an ancient field of 100 μT (after Collinson, 1984).

(see Figure 7.6). Collinson (1984) has given a good review of this work. Fuller and Cisowski (1989) agree with this general conclusion, although they believe that there was a more rapid end of the magnetic era. There have been a number of suggestions to account for this early lunar magnetic field. The simplest explanation is that is was acquired by an internal dynamo operating in a small, highly conducting, fluid Fe core. The loss of a magnetic field later than 3.1 Ga ago could be the result of solidification of the entire core, or by convection currents below the strength needed to power the dynamo.

Nakamura et al. (1974) have tentatively inferred a small lunar core within about 350 km of the lunar centre based on greatly reduced P wave velocities recorded by seismometers installed on the Moon. The origin of the lunar magnetic field is however still unresolved. If it is necessary to invoke a lunar dynamo during the early history of the moon, this would place severe constraints on geochemical models of the Moon's evolution. It would require that the Moon differentiated early in its history in order to have a metallic core, with temperatures above the melting point of iron in the lunar interior. If the lunar dynamo failed by solidification of the core, the loss of heat could be achieved by convective cooling.

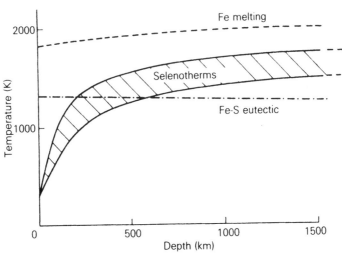

Figure 7.7 Comparison of actual lunar internal temperatures (shaded region) with the melting curve for pure iron and the eutectic curve for Fe–S. Since the actual temperatures are intermediate, any lunar core must be at least partially fluid (after Stevenson, 1983).

Tozer (1972) believes that the Moon is likely to undergo solid-state convection as soon as the temperature of the lunar interior rises above ~1000°C. Figure 7.7 shows the estimated present temperature distribution in the Moon (after Stevenson, 1983). The figure shows that the present internal temperature is well above the Fe–S eutectic, indicating that if the Moon has a core, then it must be partially fluid. The present day OC thickness is ~100 km for an initially fluid core of radius 400 km containing 5% sulphur by mass. Stevenson also showed that gravitational energy release from gradual cooling of this core is too small by about a factor of three to sustain compositionally driven convection, although gravitational energy release during core formation could possibly have provided sufficient power for dynamo generation. Because the OC is so small now, Stevenson believes that the energy available to drive the dynamo is about two orders of magnitude too low.

7.6 THE GREAT PLANETS

No attempt will be made to discuss in any detail the deep interior of the great planets, all of which have now been visited by spacecraft – a good review has been given by Stevenson (1982). Their composition is very different from that of the terrestrial planets.

They do however appear, like the Earth, to have significant magnetic fields. In the case of the terrestrial planets, core formation and the nature of the light alloying element are critical in determining their dynamic state. In Jupiter and Saturn, the dominant constituent hydrogen forms a metal at around 2–5 Mbar. The source of any intrinsic magnetic field is likely to be in these metallic hydrogen shells rather than in possible rock and/or ice cores. In the case of Saturn, the angle between the dipole and rotation axis is essentially zero, in contrast to both the Earth and Jupiter where the dipole tilt angles are ~10°. The case of Uranus is even more complex. To a first approximation the field is that of a tilted dipole, but at an angle of 60° from the rotation axis. Higher order terms contribute as much as 70% of the dipole terms and can be represented by displacing the dipole by about one third of the planet's radius from the centre away from the sunlit side. Thus the surface field is much weaker on the sunlit side (<10 μT) than on the dark side (<110 μT).

When the Voyager 2 spacecraft approached Neptune in August 1989, it was predicted that its magnetic field would resemble the 'well-behaved' fields of Mercury, the Earth, Jupiter and Saturn. Instead it was more like Uranus magnetically – forcing us to rethink how magnetic fields are generated within the interiors of planets. The magnetic field of Neptune can be represented by an offset tilted dispole, displaced from the centre by over one half the planet's radius and inclined at 47° with respect to the rotation axis. This gives it the widest range of field strength at its surface of any planet, ranging from 6 μT to 120 μT. It seemed natural to link the strangeness of Uranus' magnetic field to the planet's unique orientation. Uranus is tilted 90° – its poles are in the planes of most planets' equators. It has been suggested that this configuration is the result of a tremendous offcentre impact from another planet-sized body late in the formation of the solar system, which also disrupted the mechanism that generated the magnetic field. There is no sign, however, that any cataclysmic impact disrupted Neptune. The pictures and measurements sent back by Voyager 2 have shown that the outer solar system is even more complex and wonderful than we could have imagined.

References

Agee, C.B., (1990). A new look at differentiation of the Earth from melting experiments on the Allende meteorite. *Nature* **346**, 834.

Ahrens, T.J., (1979). Equations of state of iron sulphide and constraints on the sulphur content of the Earth. *J. Geophys. Res.* **84**, 985.

Ahrens, T.J., (1987). Shock wave techniques for geophysics and planetary physics, in: *Methods of Experimental Physics* **24**, Part A, Acad, Press.

Akogi, M., Ito, E., and Navrotsky, A.J., (1989). Olivine modified spinel-spinel transitions in the system $Mg_2SiO_4-Fe_2SiO_4$: calometric measurements, thermochemical calculation and geophysical application. *J. Geophys. Res.* **94**, 15671.

Albritton, Jr. C.C., (1989). *Catastrophic Episodes in Earth History*. Chapman and Hall.

Allègre, C.J., Dupré, B., and Brévart, O., (1982). Chemical aspects of the formation of the core. *Phil. Trans. Roy. Soc. London A*, **306**, 49.

Alvarez, L.W., Alvarez, W., Asaro, F., and Michel, H.V., (1980). Extraterrestrial cause for the Cretaceous/Tertiary extinction. *Science* **208**, 1095.

Anderson, D.L., (1989). Composition of the Earth. *Science* **243**, 367.

Anderson, D.L. and Bass, J.D., (1986). Transition region of the Earth's upper mantle. *Nature* **320**, 321.

Anderson, W.W., Svendsen, B., and Ahrens, T.J., (1989). Phase relations in iron-rich systems and implications for the Earth's core. *Phys. Earth Planet. Int.* **55**, 208.

Babcock, H.W., (1947). Zeeman effect in stellar spectra. *Astrophys. J.* **105**, 105.

Bell, P.M., and Mao, H.K., (1979). Absolute pressure measurements and their comparison with the ruby fluorescence (R_1) pressure scale to 1.5 Mbar. *Carnegie Inst. Washington, Year Book* 78, 665.

Bell, P.M., Xu, J., and Mao, H.K., (1986). Static compression of gold and copper and calibration of the ruby-pressure scale to pressures to 1.8 Megabars (Static. RNO) in, *Shock Waves in Condensed Matter*, (ed. Y.M. Gupta), Plenum Publ. Co.

Benz, W., Slattery, W.L., and Cameron, A.G.W., (1988). Collisional stripping of Mercury's mantle. *Icarus* **74**, 516.

Bercovici, D., Schubert, G., and Glatzmaier, G.A. (1989a). Influence of heating mode on three dimensional mantle convection. *Geophys. Res. Lett.* **16**, 617.

Bercovici, D., Schubert, G., and Glatzmaier, G.A., (1989b). Three-dimensional spherical models of convection in the Earth's mantle. *Science* **244**, 950.

Birch, F., (1952). Elasticity and constitution of the Earth's interior. *J. Geophys. Res.* **57**, 227.

Birch, F., (1968). On the possibility of large changes in the Earth's volume. *Phys. Earth Planet. Int.* **1**, 141.

Blackett, P.M.S., (1947). The magnetic field of massive rotating bodies. *Nature*, **159**, 658.

Blackett, P.M.S., (1952). A negative experiment relating to magnetism and the Earth's rotation. *Phil. Trans. Roy. Soc.* A 245, 309.

Bloxham, J., (1986). The expulsion of magnetic flux from the Earth's core. *Geophys. J.* **87**, 669.

Bloxham, J., and Gubbins, D., (1985). The secular variation of the Earth's magnetic field. *Nature*, **317**, 777.

Bloxham, J., and Gubbins, D., (1986). Geomagnetic field analysis – IV. Testing the frozen flux hypothesis. *Geophys. J.* **84**, 139.

Bloxham, J., and Gubbins, D., (1987). Thermal core-mantle interactions. *Nature*, **325**, 511.

Boehler, R., (1986). The phase diagram of iron to 430 kbar. *Geophys. Res. Lett.* **13**, 1153.

Boehler, R., von Bargen, N., and Chopelas, A., (1990). Melting, thermal expansion and phase transitions of iron at high pressures. *J. Geophys. Res.* **95**, 21, 731.

Bolt, B.A., (1976). *Nuclear Explosions and Earthquakes*. W.H. Freeman and Company.

Bolt, B.A., (1982). *Inside the Earth*. W.H. Freeman and Co.

Braginskii, S.I., (1963). Structure of the F layer and reasons for convection in the Earth's core. *Dokl. Akad. Nauk. SSR* **149**, 134.

Breneman, H.H., and Stone, E.C., (1985). Solar coronal and photospheric abundances from solar energetic particle measurements. *Astrophys. J.* **299**, L257.

Brett, R., (1976). The current status of speculation on the composition of the core of the Earth. *Rev. Geophys. Space Phys.* **14**, 375.

Brett, R., (1984). Chemical equilibrium of the Earth's core and upper mantle. *Geochim. Cosmochim. Acta* **48**, 1183.

Brown, J.M., (1986). Interpretation of the D″ zone at the base of the mantle: dependence on assumed values of thermal conductivity. *Geophys. Res. Lett.* **13**, 1509.

Brown, J.M., Ahrens, T.J., and Shampine, D., (1984). Hugoniot data for pyrrhotite and the Earth's core. *J. Geophys. Res.* **89**, 6041.

Bullard, E.C., (1949a). The magnetic field within the Earth. *Proc. Roy. Soc. A*, **197**, 433.

Bullard, E.C., (1949b). Electromagnetic induction in a rotating sphere. *Proc. Roy. Soc. A.* **199**, 413.

Bullard, E.C., (1968). Reversals of the Earth's magnetic field. *Phil. Trans. Roy. Soc.* **A263**, 481.

Bullen, K.E., (1953). The rigidity of the Earth's inner core. *Ann. Geofis.* **6**, 1.

Bullen, K.E., (1954). *Seismology*. Methuen, London.

Bullen, K.E., (1963, 1984). *An Introduction to the Theory of Seismology*. Camb. Univ. Press, 1963 4th Edition (with Bolt, B.A., 1984).

Bullen, K.E., (1975). *The Earth's Density*. Chapman and Hall.

Bullen, K.E., and Haddon, R.A.W., (1970). Evidence from seismology and related sources on the Earth's present internal structure. *Phys. Earth Planet. Int.* **2**, 342.

Busse, F.H., (1970). Thermal instabilities in rapidly rotating systems. *J. Fluid Mech.* **44**, 441.

Cameron, A.G.W., and Ward, W.R., (1976). The origin of the Moon. *Lunar Sci.* **7**, 120.

Campbell, I.H., and Griffiths, R.W., (1990). Implications of mantle plume structure for the evolution of flood basalts. *Earth Planet. Sci. Lett.* **99**, 79.

Chapman, C.R. (1989). Snowbird II: Global catastrophes EΘ S. *Trans. Amer. Geophys. Un.* **70**, 217.

Chen, J.H., and Wasserburg, G.J., (1986). Formation ages and evolution of Shergotty and its parent from U-Th-Pb systematics. *Geochim. Cosmochim. Acta* **50**, 955.

Christensen, U.R., and Yuen, D.A., (1984). The interaction of a subducting lithospheric slab with a chemical or phase boundary. *J. Geophys. Res.* **89**, 4389.

Cisowski, S.M., Collinson, D.W., Runcorn, S.K., Stephenson, A., and Fuller, F.M., (1983). A review of lunar palaeointensity data and implications for origin of lunar magnetism. *Proc. 13th Lunar Planet. Sci. Conf. J. Geophys. Res.* **88**, A691.

Coe, R.S., and Prevot, M., (1989). Evidence suggesting extremely rapid field variation during a geomagnetic reversal. *Earth Planet Sci. Lett.* **92**, 292.

Collinson, D.W., (1984). On the existence of magnetic fields on the Moon between 3.6 Ga ago and the present. *Phys. Earth Planet. Int.* **34**, 102.

Cook, A.H., (1980). *Interiors of the Planets.* Camb. Univ. Press.

Courtillot, V., Fraud, G., Maluski, H., van Damme, M.D., Moreau, M.G., and Besse, J., (1988). Deccan flood basalts and the Cretaceous/Tertiary boundary. *Nature*, **333**, 843.

Cowling, T.G., (1934). The magnetic field of sunspots. *Mon. Not. Roy. Astr. Soc.* **94**, 39.

Cox, A., (1968). Length of geomagnetic polarity intervals. *J. Geophys. Res.* **73**, 3247.

Cox, A., (1969). Geomagnetic reversals. *Science*, **163**, 237.

Cox, P.A., (1989). *The Elements, their origin, abundance and distribution.* Oxford Univ. Press.

Creager, K.C., and Jordan, T.H., (1986). Aspherical structure of the core-mantle boundary from PKP travel times. *Geophys. Res. Lett.* **13**, 1497.

Crocket, J.H., Officer, C.B., Wezel, F.C., and Johnson, G.D., (1988). Distribution of noble metals across the Cretaceous-Tertiary boundary at Gubbio, Italy: Iridium variation as a constraint on the duration and nature of Cretaceous-Tertiary events. *Geology*, **16**, 77.

Davies, G.F., (1985). Heat deposition and retention in a solid planet growing by impacts. *Icarus* **63**, 45.

Davies, G.F., and Gurnis, M., (1986). Interaction of mantle dregs with convection: lateral heterogeneity at the core-mantle boundary. *Geophys. Res. Lett.* **13**, 1517.

Dolginov, Sh. Sh. (1987). What have we learned about the Martian magnetic field? *Earth, Moon and Planets.* **37**, 17.

Doornbos, D.J., and Hilton, T., (1989). Models of the core-mantle boundary and the travel times of internally reflected core phases. *J. Geophys. Res.* **94**, 15, 741.

Durrani, S.A., and Khan, H.A., (1971). Ivory coast microtektites: fission track age and geomagnetic reversals. *Nature*, **232**, 320.

Dziewonski, A.M., (1984). Mapping the lower mantle: determination of lateral heterogeneity in P velocity up to degree and order 6. *J. Geophys. Res.* **89**, 5929.

Dziewonski, A.M., and Anderson, D.L., (1981). Preliminary reference Earth model. *Phys. Earth Planet. Int.* **25**, 297.

Dziewonski, A.M., and Anderson, D.L., (1984). Seismic tomography of the Earth's interior. *Amer. Sci.* Sept.-Oct, 483.

Dziewonski, A.M., and Gilbert, F., (1971). Solidity of the inner core of the Earth inferred from normal mode observations. *Nature*, **234**, 465.

Dziewonski, A.M., and Woodhouse, J.H., (1987). Global images of the Earth's interior. *Science* **236**, 37.

Elsasser, W.M., (1946a). Induction effects in terrestrial magnetism, Part I Theory. *Phys. Rev.*, **69**, 106.

Elsasser, W.M., (1946b). Induction effects in terrestrial magnetism, Part II The secular variation. *Phys. Rev.* **70**, 202.

Elsasser, W.M., (1947). Induction effects in terrestrial magnetism. Part III Electric modes. *Phys. Rev.* **72**, 821.

Elsasser, W.M., (1963). Early history of the Earth, in: *Earth Science and Meteoritics*, (eds. J. Geiss and E.D. Goldberg) North-Holland, Amsterdam.

Engdahl, E.R., Flinn, E.A., and Romney, C.F., (1970). Seismic waves reflected from the Earth's inner core. *Nature*, **228**, 852.

Eugster, O., (1989). History of meteorites from the Moon collected in Antarctica. *Science*, **245**, 1197.

Flaser, F.M., and Birch, F., (1973). Energetics of core formation. *J. Geophys. Res.* **78**, 6101.

Ganguly, J., and Kennedy, G.C., (1977). Solubility of K in Fe−S liquid silicate-K-(Fe−S)^{liq} equilibria and their implications. *Earth Planet. Sci. Lett.* **35**, 411.

Giardini, D., Li, X-D., and Woodhouse, J.H., (1987). Three dimensional structure of the Earth from splitting in free-oscillation spectra. *Nature*, **325**, 405.

Glass, B.P., and Heezen, B.C., (1967). Tektites and geomagnetic reversals. *Sci. Amer.* 217 (7), 32.

Glass, B.P., and Zwart, P.A., (1979). The Ivory coast microtektite strewn field: new data. *Earth Planet. Sci.: Lett.* **43**, 336.

Glatzmaier, G.A., Schubert, G., and Bercovici, D., (1990). Chaotic, subductionlike downflows in a spherical model of convection in the Earth's mantle. *Nature*, **347**, 274.

Gleick, J., (1988). *Chaos*. Heinemann.

Goettel, K.A., and Lewis, J.S., (1973). Comments on a paper by V.M. Oversby and A.E., Ringwood *(Earth Planet Sci. Lett.* **14**, 345, 1972). *Earth Planet. Sci. Lett.* **18**, 148.

Gray, C.M., and Compston, W., (1974). Excess ^{26}Mg in the Allende meteorite. *Nature* **251**, 495.

Grieve, R.A.F., Sharpton, V.L., Goodacre, A.K., and Garvin, J.B., (1985). A perspective on the evidence for periodic cometary impacts on Earth. *Earth Planet. Sci. Lett.* **76**, 1.

Griffiths, R.W., and Campbell, I.H., (1990). Stirring and structure in mantle starting plumes. *Earth Planet. Sci. Lett.* **99**, 66.

Gubbins, D., (1981). Planetary magnetism and the thermal evolution of planetary cores; in: Evolution of the Earth. *Geodyn. Ser.* (eds. R.J. O'Connell and W.S. Fyfe), *Amer. Geophys. Un.*, Washington, DC.

Gubbins, D., and Bloxham, J., (1987). Morphology of the geomagnetic field and implications for the geodynamo. *Nature*, **325**, 509.

Gubbins, D., Masters, T.G., and Jacobs, J.A., (1979). Thermal evolution of the Earth's core. *Geophys. J.* **59**, 57.

Gubbins, D., and Richards, M., (1986). Coupling of the core dynamo and mantle: thermal or topographic? *Geophys. Res. Lett.* **13**, 1521.

Gutenberg, B., (1959). *Physics of the Earth's Interior*. Acad. Press.

Hager, B.H., Clayton, R.W., Richards, M.A., Comer, R.P., and Dziewonski, A.M., (1985). Lower mantle heterogeneity, dynamic topography and the geoid. *Nature*, **313**, 541.

Hale, C.J., (1987). Palaeomagnetic data suggest link between Archaean-Proterozoic boundary and inner-core nucleation. *Nature*, **329**, 233.

Hall, H.T., and Murthy, V.R., (1971). The early chemical history of the Earth: some critical elemental fractionations. *Earth Planet. Sci. Lett.* **11**, 239.

Harland, W.B., Cox, A.V., Llewellyn, P.G., Pickton, C.A.G., Smith, A.G., and Walters, R., (1990). *A geologic time scale*. Camb. Univ. Press.

Hayashi, C., Nakazawa, H., and Mizuno, H., (1979). Earth's melting due to the blanketing effect of the primordial dense atmosphere. *Earth Planet. Sci. Lett.* **43**, 22.

Head, J.W., and Crumpler, L.S., (1990). Venus geology and tectonics: hotspot and crustal spreading models and questions for the Magellan mission. *Nature*, **346**, 525.

Heinz, D.L., and Jeanloz, R., (1987). Measurement of the melting curve of $Mg_{0.9}Fe_{0.1}SiO_3$ perovskite at lower mantle conditions and its geophysical implications. *J. Geophys. Res.* **92**, 11437.

Hide, R. and Palmer, T.N., (1982). Generalization of Cowling's theorem. *Geophys. Astrophys. Fluid. Dyn.* **19**, 301.

Higgins, G.H., and Kennedy, G.C., (1971). The adiabatic gradient and the melting point gradient in the core of the Earth. *J. Geophys. Res.* **76**, 1870.

Huang, J., and Turcotte, D.L., (1990). Are earthquakes an example of deterministic chaos? *Geophys. Res. Lett.* **17**, 223.

Hutcheon, I.D., and Hutchison, R., (1989). Evidence from the Samarkona ordinary chondrite for ^{26}Al heating of small planets. *Nature*, **337**, 238.

Inoue, H., Fukao, Y., Tanabe, K., and Ogata, Y., (1990). Whole mantle P-wave travel-time tomography. *Phys. Earth Planet. Int.* **59**, 294.

Irving, E., and Pullaiah, G., (1976). Reversals of the geomagnetic field, magneto-stratigraphy and relative magnitude of palaeosecular variation in the Phanerozoic. *Earth Sci. Rev.* **12**, 35.

Ito, E., and Takahashi, E., (1989). Postspinel transformations in the system $Mg_2SiO_4-Fe_2SiO_4$ and some geophysical implications. *J. Geophys. Res.* **94**, 10637.

Ito, E., Akogi, M., Topor, L., and Navrotsky, A., (1990). Negative pressure-temperature slopes for reactions forming $MgSiO_3$ perovskite from calorimetry. *Science*, **249**, 1275.

Jacobs, J.A., (1953). The Earth's inner core. *Nature*, **172**, 297.

Jacobs, J.A., (1974). *A Textbook on Geonomy*. Adam Hilger, Bristol.

Jacobs, J.A., (1980). The evolution of the Earth's core and the geodynamo, in: *Physics of the Earth's Interior, Proc. Inter. School Phys.* Enrico Fermi, Course LXXVIII, (eds. A.M. Dziewonski and E. Boschi), North Holland Publ. Co.

Jacobs, J.A., (1984). *Reversals of the Earth's Magnetic Field*. Adam Hilger, Bristol.

Jacobs, J.A., (1987). *The Earth's Core*. 2nd Edition. Acad. Press.

Jeanloz, R., (1979). Properties of iron at high pressures and the state of the core. *J. Geophys. Res.* **84**, 6059.

Jeanloz, R., (1990). The nature of the Earth's core. *Ann. Rev. Earth Planet. Sci.* **18**, 357.

Jeanloz, R., and Ahrens, T.J., (1980). Equation of state of FeO and CaO. *Geophys. J.* **62**, 505.

Jeanloz, R., and Morris, S., (1986). Temperature distribution in the crust and mantle. *Ann. Rev. Earth Planet. Sci.* **14**, 377.

Jeanloz, R., and Wenk, J-R., (1988). Convection and anisotrophy of the inner core. *Geophys. Res. Lett.* **15**, 72.

Jephcoat, A., and Olson, P., (1987). Is the inner core of the Earth pure iron? *Nature*, **325**, 332.

Jones, G.M., (1977). Thermal interaction of the core and the mantle and long term behaviour of the geomagnetic field. *J. Geophys. Res.* **82**, 1703.

Kaula, W.M., (1980). The beginning of the Earth's thermal evolution, in: *The Continental Crust and its Mineral Deposits* (ed. D.W. Strangeway). Geol. Soc. Canada Spec. Paper 20.

Kellogg, L.H., and Turcotte, D.L., (1990). Mixing and the distributions of heterogeneities in a chaotically convecting mantle. *J. Geophys. Res.* **95**, 421.

Kennedy, G., and Higgins, G.H., (1973). The core paradox. *J. Geophys. Res.* **78**, 900.

Kincaid, C., and Olson, P., (1987). An experimental study of subduction and slab migration. *J. Geophys. Res.* **92**, 13832.

Knittle, E., and Jeanloz, R., (1986). High pressure metallization of FeO and implications for the Earth's core. *Geophys. Res. Lett.* **13**, 1541.

Knittle, E. Jeanloz, R., (1987). Synthesis and equation of state of (Mg, Fe)SiO₃ perovskite to over 100 GPa. *Science*, **235**, 668.

Knittle, E., and Jeanloz, R., (1989a). Melting curve of (Mg, Fe) SiO₃ perovskite to 96 GPa; evidence for a structured transition lower mantle melts. *Geophys. Res. Lett.* **16**, 421.

Knittle, E., and Jeanloz, R., (1989b). Simulating the core-mantle boundary: an experimental study of high-pressure reactions between silicates and liquid iron. *Geophys. Res. Lett.* **16**, 609.

Knittle, E., and Jeanloz, R., (1991a). The high pressure phase diagram of Fe₀.₉₄O, a possible constituent of the Earth's core. *J. Geophys. Res.*

Knittle, E., and Jeanloz, R., (1991b). Earth's core–mantle boundary: results of experiments at high pressures and temperatures. *Science*, **251**, 1438.

Knopoff, L., (1964). Q. *Rev. Geophys.* **2**, 625.

Kraft, A., Stiller, H., and Vollstadt, H., (1982). The monosulphid solution in the Fe-Ni-S system: relationship to the Earth's core on the basis of experimental high-pressure investigations. *Phys. Earth Planet. Int.* **27**, 255.

Krause, F., and Rädler, K.H., (1980). *Mean Field Magnetohydrodynamics and Dynamo Theory*. Oxford: Pergamon.

Lachenbruch, A.H., (1976). Dynamics of a passive spreading centre. *J., Geophys. Res.* **81**, 1883.

Laj, C., Guitton, S., Kissel, C., and Mazaud, A., (1988). Complex behaviour of the geomagnetic field during three successive polarity reversals, 11–12 my B.P. *J. Geophys. Res.* **93**, 11655.

Larmor, J., (1919). How could a rotating body such as the sun become a magnet? *Rept. Brit. Assoc.* 159.

Laskar, J., (1989). A numerical experiment on the chaotic behaviour of the solar system. *Nature*, **338**, 237.

Lay, T., and Helmberger, D.V., (1983). A lower mantle S wave triplication and the shear velocity structure of D". *Geophys. J.* **75**, 799.

Lay, T., and Young, C.J., (1990). The stably-stratified core revisited. *Geophys. Res. Lett.* **17**, 2001.

Lay, T., Ahrens, T.J., Olson, P., Smyth, J., and Loper, D., (1990). Studies of the Earth's deep interior: goals and trends. *Physics To-Day*, Oct., p. 44.

Typhoon, Lee, Papanastassiou, D.A., and Wasserburg, G.J., (1976) Demonstration of ²⁶Mg excess in Allende and evidence for ²⁶Al. *Geophys. Res. Lett.* **3**, 109.

Lehmann, I. (1935). P' *Publ. Bur. Cent. Seism. Int. Ser.* **14**, 3.

Lewis, J.S., (1971). Consequences of the presence of sulphur in the core of the Earth. *Earth Planet. Sci. Lett.* **11**, 130.

Loper, D.E., (1978). Some thermal consequences of a

gravitationally powered dynamo. *J. Geophys. Res.* **83**, 5961.

Loper, D.E., (1985). A simple model of whole-mantle convection. *J. Geophys. Res.* **90**, 1809.

Loper, D.E., and Eltayeb, I.A., (1986). On the stability of the D″ layer. *Geophys. Astrophys. Fluid Dyn.* **36**, 229.

Loper, D.E., and McCartney, K., (1986). Mantle plumes and the periodicity of magnetic field reversals. *Geophys. Res. Lett.* **13**, 1525.

Loper, D.E., and Roberts, P.H., (1978). On the motion of an iron-alloy core containing a slurry 1. General theory. *Geophys. Astrophys. Fluid Dyn.* **9**, 289.

Loper, D.E., and Roberts, P.H., (1981). A study of conditions at the inner core boundary of the Earth. *Phys. Earth Planet. Int.* **24**, 302.

Loper, D.E., and Roberts, P.H., (1983). Compositional convection and the gravitationally powered dynamo, in: *Stellar and Planetary Magnetism* (ed. A.M. Soward). Gordon & Breach, London.

Loper, D.E., and Stacey, F.D., (1983). The dynamical and thermal structure of deep mantle plumes. *Phys. Earth Planet. Int.* **3**, 304.

Loper, D.E., McCartney, K., and Buzyna, G., (1988). A model of correlated episodicity in mgnetic field reversals, climate and mass extinction. *J. Geol.* **96**, 1.

Lowrie, W., and Alvarez, W., (1977). Upper Cretaceous-Palaeocene magnetic stratigraphy at Gubbio, Italy II. Upper Cretaceous magnetic stratigraphy. *Bull. Geol. Soc. Amer.* **88**, 874.

Lutz, T.M., (1985). The magnetic record is not periodic. *Nature*, **317**, 404.

Machetel, P., and Weber, P., (1991). Intermittent layered convection in a model mantle with an endothermic phase change at 670 km. *Nature*, **350**, 55.

Mankinen, E.A., Prevot, M., Gromme, C.S., and Coe, R.S., (1985). The Steens Mountain (Oregon) geomagnetic polarity transitions, 1. Directional history, duration of episodes, and rock magnetism. *J. Geophys. Res.* **90**, 10393.

Mao, H.K., and Bell, P.M., (1976). Compressibility and X-ray diffraction of the epsilon phase of metallic iron (ε-Fe) and periclase (MgO) to 0.9 Mbar pressure with bearing on the mantle core boundary. *Carnegie Inst. Washington Year Book* 75, 509.

Mao, H.K., and Bell, P.M., (1978). High-pressure physics: sustained static generation of 1.36 to 1.72 Megabars. *Science*, **200**, 1145.

Mao, H.K., Goettel, K.A., and Bell, P.M., (1985). Ultra high-pressure experiments at pressures exceeding 2 megabars, in: *Solid State Physics Under Pressure: Recent Advance with anvil devices*, (ed.

S. Minomura). D. Reidel Publ. Co., 1985.

Masters, G., (1979). Observational constraints on the chemical and thermal structure of the Earth's deep interior. *Geophys. J.* **57**, 507.

Matsui, T., and Abe, Y., (1986). Formation of a 'magma ocean' on the terrestrial planets due to the blanketing effect of an impact induced atmosphere. *Earth, Moon and Planets*, **34**, 223.

McCammon, C.A., Ringwood, A.E., and Jackson, I., (1983). Thermodynamics of the system Fe–FeO–MgO at high pressure and temperature and a model for the formation of the Earth's core. *Geophys. J.* **72**, 577.

McElhinny, M.W., (1971). Geomagnetic reversals during the Phanerozoic. *Science*, **172**, 157.

McFadden, P.L., and Merrill, R.T., (1984). Lower mantle convection and geomagnetism. *J. Geophys. Res.* **89**, 3354.

McSween, Jr. H.Y., (1984). SNC meteorites: are they Martian rocks? *Geology*, **12**, 3.

de Menocal, P.B., Ruddiman, W.F., and Kent, D.V., (1990). Depth of postdepositional acquisition in deep-sea sediments: a case study of the Brunhes-Matuyama reversal and oxygen isotope stage 19.1. Earth Planet. *Sci. Lett.* **99**, 1.

Merrill, R.T. and McFadden, P.L., (1990). Paleomagnetism and the nature of the Geodynamo. *Science*, **248**, 345.

Moffatt, H.K., (1970). Turbulent dynamo action at low magnetic Reynolds number. *J. Fluid Mech.* **41**, 435.

Moffatt, H.K., (1978). *Magnetic field generation in electrically conducting fluids*. Camb. Univ. Press.

Morelli, A., Dziewonski, A.M., and Woodhouse, J.H., (1986). Anisotropy of the inner core inferred from PKIKP travel times. *Geophys. Res. Lett.* **13**, 1545.

Morelli, A., and Dziewonski, A.M., (1987). Topography of the core-mantle boundary and lateral homogeneity of the liquid core. *Nature*, **325**, 678.

Morgan, W.J., (1971). Convection plumes in the lower mantle. *Nature*, **230**, 42.

Morgan, W.J., (1972). Plate motions and deep mantle convection. *Geol. Soc. Amer. Mem.* **132**, 7.

Muller, R.A., and Morris, D.E., (1986). Geomagnetic reversals from impacts on the Earth. *Geophys. Res. Lett.* **13**, 1177.

Murrell, M.T., and Burnett, D.S., (1986). Partitioning of K, U, and Th between sulphide and silicate liquids: implications for radioactive heating of planetary cores. *J. Geophys. Res.* **91**, 8126.

Murthy, V.R., and Hall, H.T., (1970). The chemical composition of the Earth's core; possibility of sulphur in the core. *Phys. Earth*

Planet. Int. **2**, 276.

Nakamura, Y., Latham, G., Lammlein, D., Ewing, M., Dunnelier, F., and Dorman, J., (1974). Deep lunar interior inferred from recent seismic data. *Geophys. Res. Lett.* **1**, 137.

Newson, H.E., and Taylor, S.R., (1989). Geochemical implications of the formation of the Moon by a single giant impact. *Nature*, **338**, 29.

Ninomiya, T., (1978). Theory of melting, dislocation model. *J. Phys. Soc. Japan* **44**, 263.

Ohtani, E., and Ringwood, A.E., (1984a). Composition of the core I. solubility of oxygen in molten iron at high temperatures. *Earth Planet. Sci. Lett.* **71**, 85.

Ohtani, E., Ringwood, A.E., and Hibberson, W., (1984b). Composition of the core II. effect of high pressure on solubility of FeO in molten iron. *Earth Planet. Sci. Lett.* **71**, 94.

Olson, P., Schubert, G., and Anderson, C., (1987). Plume formation in the D'' layer and the roughness of the core–mantle boundary. *Nature*, **327**, 409.

Olson, P., Silver, P.G., and Carlson, R.W., (1990). The large-scale structure of convection in the Earth's mantle. *Nature*, **344**, 209.

Opdyke, N.D., Glass, B., Hays, J.D., and Foster, J., (1966). Palaeomagnetic study of Antarctic deep sea cores. *Science*, **154**, 349.

O'Reilly, W., (1984). *Rock and mineral magnetism.* Blackie.

Oversby, V.M., and Ringwood, A.E., (1971). Time of formation of the Earth's core. *Nature*, **234**, 463.

Oversby, V.M., and Ringwood, A.E., (1972). Potassium distribution between metal and silicate and its bearing on the occurrence of potassium in the Earth's core. *Earth Planet. Sci. Lett.* **14**, 345.

Ozima, M., Podosek, F.A., and Igarashi, G., (1985). Terrestrial xenon isotope constraints on the early history of the Earth. *Nature*, **315**, 471.

Parker, R.L., (1972). Inverse theory with grossly inadequate data. *Geophys. J.* **29**, 123.

Pekeris, C.L., Alterman, Z., and Jarosch, H., (1961). Rotational multiplets in the spectrum of the Earth. *Phys. Rev.* **122**, 1692.

Poirier, J.P., (1986). Dislocation mediated melting of iron and the temperature of the Earth's core. *Geophys. J.* **85**, 315.

Press, F., (1968). Earth models obtained by Monte Carlo inversion. *J. Geophys. Res.* **73**, 5223.

Prévot, M., Mankinen, E.A., Coe, R.S., and Grommé C.S., (1985). The Steens Mountain (Oregon) geomagnetic polarity transition 2. Field intensity variations and discussion of reversal models. *J. Geophys. Res.* **90**, 10417.

Prevot, M., Derder, M.E., McWilliams, M., and Thompson, J., (1990). Intensity of the Earth's magnetic field: evidence for a Mesozoic dipole low. *Earth Planet Sci. Lett.* **97**, 129.

Rampino, M.R., and Stothers, R.B., (1984). Terrestrial mass extinctions, cometry impacts and the sun's motion perpendicular to the galactic plane. *Nature*, **308**, 709.

Raup, D.M., (1985). Magnetic reversals and mass extinctions. *Nature*, **314**, 341.

Raup, D.M., and Sepkowski, J.J., (1986). Periodic extinction of families and genera. *Science*, **231**, 833.

Raup, D.M., and Sepkowski, J.J., (1988). Testing for periodicity of extinction. *Science*, **241**, 94.

Richter, F.M., and McKenzie, D.P., (1978). Simple plate models of mantle convection. *J. Geophys.* **44**, 441.

Riedler, W., *et al.*, (1989). Magnetic field near Mars: first results. *Nature*, **341**, 604.

Ringwood, A.E., (1960). Some aspects of the thermal evolution of the Earth. *Geochim. Cosmochim. Acta* **20**, 241.

Ringwood, A.E., (1966). Chemical evolution of the terrestrial planets. *Geochim. Cosmochim. Acta*, **30**, 41.

Ringwood, A.E., (1978). Composition of the core and implication for origin of the Earth. *Geochem. J.* **11**, 111.

Ringwood, A.E., (1979). *Origin of the Earth and Moon*. Springer-Verlag.

Ringwood, A.E., (1989). Flaws in the giant impact hypothesis of lunar origin. *Earth Planet. Sci. Lett.* **95**, 208.

Ringwood, A.E., Kato, T., Hibberson, W., and Ware, N., (1990). High pressure geochemistry of Cr, V and Mn and implications for the origin of the Moon. *Nature*, **347**, 174.

Rocchia, R., Boclet, D., Bonté, Ph., Jéhanno, C., Yan Chen, Courtillot, V., Mary, C., and Wezel, F., (1990). The Cretaceous-Tertiary boundary at Gubbio revisited: vertical extent of the Ir anomaly. *Earth Planet. Sci. Lett.*, **99**, 206.

Russell, C.T., (1987). Planetary Magnetism, in: *Geomagnetism Volume 2* (ed. J.A. Jacobs), Acad. Press.

Ryder, G., (1990). *Lunar samples, lunar accretion and the early bombardment of the Moon*. EOS 313, (March 6).

Safronov, V.S., (1978). The heating of the Earth during its formation. *Icarus* **33**, 1.

Saunders, R.S., Pettengill, G.H., Arvidson, R.E., Sjogren, W.L., Johnson, W.T.K., and Pieri, L., (1990). The Magellan Venus radar mapping mission. *J. Geophys. Res.* **95**, 8339.

Sayers, C.M., (1990). The crystal structure of iron in the Earth's inner core. *Geophys. J. Int.* **103**, 285.

Schneider. D.A., and Kent, D.V., (1990). Ivory coast microtektites and geomagnetic reversals. *Geophys. Res. Lett.* **17**, 163.

Schubert, G., (1979). Subsolidus convection in the mantles of terrestrial planets. *Ann. Rev. Earth Planet. Sci.* **7**, 289.

Schubert, G., Yuen, D.A., and Turcotte, D.L., (1975). Role of phase transitions in a dynamic mantle. *Geophys. J.* **42**, 705.

Schubert, G. and Spohn, T., (1990). Thermal history of Mars and the sulfur content of its core. *J. Geophys. Res.* **95**, 14, 095.

Schubert, G., Bercovici, D., and Glatzmaier, G.A., (1990). Mantle dynamics in Mars and Venus: influence of an immobile lithosphere on three-dimensional mantle convection. *J. Geophys, Res.* **95**, 14, 105.

Seitz, M.G., and Kushiro, I., (1974). Melting relations of the Allende meteorite. *Science*, **183**, 954.

Shaw, G.H., (1978). Effects of core formation. *Phys. Earth Planet. Int.* **16**, 361.

Shaw, H.F., and Wasserburg, G.J., (1982). Age and provenance of the target materials for tektites and possible impactites as inferred from Sm-Nd and Rb-Sr systematics. *Earth Planet. Sci. Lett.* **60**, 155.

Shaw, J., (1975). Strong geomagnetic fields during a single Icelandic polarity transition. *Geophys. J.* **40**, 345.

Shearer, P.M., Toy, K.M., and Orcutt, J.A., (1988). Axi-symmetric Earth models and inner-core anisotropy. *Nature*, **333**, 228.

Shearer, P., and Masters, G., (1990). The density and shear velocity contrast at the inner core boundary. *Geophys. J. Int.* **102**, 491.

Shearer, P.M., and Toy, K.M., (1991). PKP (BC) versus PKP (DF) differential travel times and aspherical structure in the Earth's inner core. *J. Geophys. Res.*, **96**, 2233.

Sheridan, R.E., (1983). Phenomena of pulsation tectonics related to the breaking up of the eastern North American continental margin. *Tectonophys.* **94**, 169.

Silver, P.G., Carlson, R.W., and Olson, P., (1988). Deep slabs, geochemical heterogeneity, and the large scale structure of mantle convection: investigation of an enduring paradox. *Ann. Rev. Earth Planet. Sci.* **16**, 477.

Solomatov, V.S., and Zharkov, V.N., (1990). The thermal regime of Venus. *Icarus*, **84**, 280.

Somerville, M.R., and Ahrens, T.J., (1980). Shock compression of $K Fe S_2$ and the question of potassium in the core. *J. Geophys. Res.* **85**, 7016.

Souriau, A., and Souriau, M., (1989). Ellipticity and density at the inner core boundary from sub-critical PkiKP and PcP data.

Geophys. J. Int. **98**, 39.

Spohn, T., (1991). Mantle differentiation and thermal evolution of Mars, Mercury and Venus. *Icarus*, **90**, 222.

Stacey, F.D., (1972). Physical properties of the Earth's core. *Geophys. Surv.* **1**, 99.

Stacey, F.D., (1975). Thermal regime of the Earth's interior. *Nature*, **255**, 44.

Stacey, F.D., and Irvine, R.D., (1977). Theory of melting; thermodynamic basis of Lindemann law. *Aust. J. Phys.* **30**, 631.

Stacey, F.D., and Loper, D.E., (1983). The thermal boundary layer intepretation of D″ and its role as a plume source. *Phys. Earth Planet. Int.* **33**, 45.

Stevenson, D.J., (1982). Interiors of the giant planets. *Ann. Rev. Earth Planet. Sci.* **10**, 257.

Stevenson, D.J., (1983). Planetary magnetic fields. *Rep. Prog. Phys.* **46**, 555.

Stevenson, D.J., (1987). Origin of the Moon – the collision hypothesis. *Ann. Rev. Earth Planet. Sci.* **15**, 271.

Stevenson, D.J., Spohn, T., and Schubert, D.J., (1983). Magnetism and thermal evolution of the terrestrial planets. *Icarus*, **54**, 466.

Stixrude, L., and Bukowinski, M.S.T., (1990). Fundamental thermodynamic relations and silicate melting with implications for the constitution of D″ *J. Geophys. Res.* **95**, 19, 311.

Sussman, G.J., and Wisdom, J., (1988). Numerical evidence that the motion of Pluto is chaotic. *Science*, **241**, 433.

Svendsen, B., Anderson, W.W., Ahrens, T.J., and Bass, J.D., (1989). Ideal Fe–Fes, Fe–FeO phase relations in the Earth's core. *Phys. Earth Planet Int.* **55**, 154.

Tozer, D.C., (1965). Thermal history of the Earth 1. The formation of the core. *Geophys. J.* **9**, 95.

Tozer, D.C., (1972a). The present thermal state of the terrestrial planets. *Phys. Earth Planet Int.* **6**, 182.

Tozer, D.C., (1972b). The Moon's thermal state and an interpretation of the lunar electrical conductivity distribution. *The Moon*, **5**, 90.

Tozer, D.C., (1977). The thermal state and evolution of the Earth and terrestrial planets. *Sci. Prog. Oxford*, **64**, 1.

Verhoogen, J., (1961). Heat balance of the Earth's core. *Geophys. J.* **4**, 276.

Vityazev, A.V., (1973). On gravitional differentiation energy in the Earth. *Izv. Earth Phys.* **10**, 676.

Vollmer, R., (1977). Terrestrial lead isotopic evolution and formation time of the Earth's core. *Nature*, **270**, 144.

Watkins, N.D., (1968). Short period geomagnetic polarity events in deep-sea sedimentary cores. *Earth Planet Sci. Lett.* **4**, 341.

Wetherill, G.W., (1980). Formation of the terrestrial planets. *Ann. Rev. Astron. Astrophys.* **18**, 77.

Wetherill, G.W., (1985). Occurrence of giant impacts during the growth of the terrestrial planets. *Science*, **228**, 877.

Wetherill, G.W., (1989). In *Mercury* (eds. C. Chapman, and F. Vilas). Univ. Arizona Press.

Wetherill, G.W., (1990). Formation of the Earth. *Ann. Rev. Earth Planet. Sci.*, **18**, 205.

Widmer, R., Masters, G., and Gilbert, F., (1988). The spherically symmetric Earth: observational aspects and constraints on new models, EOS. *Trans. Amer. Geophys. Un.* **69**, 1310.

Williams, Q., Jeanloz, R., Bass, J., Svendsen, B., and Ahrens, T.J., (1987). The melting curve of iron to 250 gigapascals: a constraint on the temperature at Earth's centre. *Science*, **236**, 181.

Williams, Q., and Jeanloz, R., (1990). Melting relations in the iron-sulphur system at ultra-high pressures: implications for the thermal state of the Earth. *J. Geophys. Res.* **95**, 19, 299.

Williams, Q., Knittle, E., and Jeanloz, R., (1991). The high-pressure melting curve of iron: a technical discussion. *J. Geophys. Res.*, **96**, 2171.

Williamson, E.D., and Adams, L.H., (1923). Density distribution in the Earth. *J. Wash, Acad, Sci.* **13**, 413.

Wilson, J.T., (1963). A possible origin of the Hawaiian islands. *Can. J. Phys.* **41**, 863.

Wilson, J.T., (1965). Convection currents and continental drift: evidence from ocean islands suggesting movement in the Earth. *Phil. Trans. Roy. Soc. London A.* **258**, 145.

Wisdom, J., Peale, S.J., and Mignard, F., (1984). The chaotic rotation of Hyperion. *Icarus*, **58**, 137.

Wisdom, J., (1987). Urey prize lecture: chaotic dynamics in the solar system. *Icarus*, **72**, 241.

Woodhouse, J.H., and Dziewonski, A.M., (1984). Mapping the upper mantle: three dimensional modelling of Earth structure by inversion of seismic waveforms. *J. Geophys. Res.* **89**, 5933.

Xu, J., Mao, H.K., and Bell, P.M., (1986). High pressure ruby and diamond fluorescence observations at 0.21 to 0.55 terrapascal. *Science*, **232**, 1404.

Young, C.J., and Lay, T., (1987). The core-mantle boundary. *Ann. Rev. Earth Planet Sci.* **15**, 25.

Young, C.J., and Lay, T., (1989). The core shadow zone boundary and lateral variations of the *P* velocity structure of the lowermost mantle. *Phys. Earth Planet. Int.* **54**, 64.

Yuen, D.A., and Peltier, W.R., (1980). Mantle plumes and the thermal stability of the D'' layer. *Geophys. Res. Lett.* **7**, 625.

Index

Printed in the United States
By Bookmasters